AIR POLLUTION CONTROL ENGINEERING

POLLUTION ENGINEERING AND TECHNOLOGY

A Series of Reference Books and Textbooks

EDITORS

1. Energy from Solid Wastes, *Paul N. Cheremisinoff and Angelo C. Morresi*
2. Air Pollution Control and Design Handbook (in two parts), *edited by Paul N. Cheremisinoff and Richard A. Young*
3. Wastewater Renovation and Reuse, *edited by Frank M. D'Itri*
4. Water and Wastewater Treatment: Calculations for Chemical and Physical Processes, *Michael J. Humenick, Jr.*
5. Biofouling Control Procedures, *edited by Loren D. Jensen*
6. Managing the Heavy Metals on the Land, *G. W. Leeper*
7. Combustion and Incineration Processes: Applications in Environmental Engineering, *Walter R. Niessen*
8. Electrostatic Precipitation, *Sabert Oglesby, Jr. and Grady B. Nichols*
9. Benzene: Basic and Hazardous Properties, *Paul N. Cheremisinoff and Angelo C. Morresi*
10. Air Pollution Control Engineering: Basic Calculations for Particulate Collection, *William Licht*

Additional volumes in preparation

AIR POLLUTION CONTROL ENGINEERING

Basic Calculations for Particulate Collection

WILLIAM LICHT

Department of Chemical
and Nuclear Engineering
University of Cincinnati
Cincinnati, Ohio

MARCEL DEKKER, INC. New York and Basel

Library of Congress Cataloging in Publication Data

Licht, William.
Air pollution control engineering.

(Pollution engineering and technology ; 10)
Includes index.
1. Gases--Purification--Mathematical models.
2. Dust--Removal--Mathematical models. I. Title.
II. Series.
TD884.5.L52 628.5'3 79-24579
ISBN 0-8247-6846-9

MARCEL DEKKER, INC.

270 Madison Avenue, New York, New York 10016

Current printing (last digit):

10 9 8 7 6 5 4 3 2 1

PRINTED IN THE UNITED STATES OF AMERICA

PREFACE

This book is concerned with how to make some basic calculations involved in air pollution control engineering. Specifically it has to do with the cleaning of gases by the removal of particulate matter.

The Federal legislation embodied in the Clean Air Act of 1963, the Air Quality Act of 1967, and the Clean Air Amendments of 1970, has resulted in the setting of ambient air quality standards and of emission regulations by governmental agencies, on the basis of protecting public health and welfare. The enforcement of these laws by the government at local, state, and national levels has become a dominant factor in the design of present day particulate collection systems. It has become necessary to do an increasingly better job of particulate collection.

In meeting this need the engineering profession is faced with two major problems: to improve its "tools" for this kind of design work, and to teach many practicing engineers (as well as students) the specialized technical skills required. The response has involved (a) research to elucidate scientific principles related to forces acting upon suspended particles and their movement, and (b) development of more precise and detailed engineering application of these principles to analyzing, synthesizing, and optimizing the performance of devices and systems. As a result, there has accumulated a large body of technical information scattered through many scientific and engineering publications: journals, patents, government reports, trade literature, etc. There has also been an outpouring of educational programs: formal degree programs at the universities, and intensive continuing education courses given by professional societies and by the U.S. Environmental Protection Agency.

This book attempts to present the relevant body of information in a unified approach to basic principles involved and their applications. It is devoted to presentation of basic (scientific) principles and basic (engineering) calculations involved in the design of particulate collection systems. It seeks to provide this information in such a way as may be useful not only in formal courses, but also to the practicing engineer previously untrained in this area and confronted with a need to teach himself how to work in it.

This is not a book devoted to descriptions of equipment or to specific problems of application to certain industries. It is intended only to show how to do fundamental calculations which underly all kinds of collection problems. It assumes some familiarity with equipment, hardware, and control techniques.

The most important design tool is an adequate mathematical model of the collection process. This model must do two things: predict how successful the collection will be for a given design and set of operating conditions, and predict the energy required at the same time. The model can then be used to explore alternative designs or systems, to optimize a design, and to provide a basis for estimation of the cost of the system.

Construction of a mathematical model for a given type of particle collector usually involves a fairly intricate combination of underlying scientific principles. As more knowledge becomes available models may become more accurate, but also more sophisticated and complex. There is great urgency to speed up this process in order to meet the goals set by the Clean Air Act.

A presentation of the calculational tools falls naturally into two parts. The first covers the basic scientific principles. This includes the methods of describing particulate systems, an overview of the fundamental concepts of collection, and the physics of particle behavior. These topics are presented in Part I, Chapters 2 through 5.

Part II deals specifically with the modelling process. It gives the best current models for cyclones, filter, electrostatic precipitators, and scrubbers in Chapter 6 through 9. Hopefully it also shows how to proceed in building models for other devices which may be developed.

A mixture of metric (S.I.) and traditional English units is used, because this is the state of affairs today. While we are on our way to adoption and widespread use of the S.I. System, many customary values in the old system remain in use and must be understood. Where these are appropriate they are retained, without apology.

The author is indebted to a number of colleagues for ideas and fruitful discussions, especially to Prof. Wm. E. Ranz of the University of Minnesota (where the manuscript first began to develop), to Profs. J.N. Pattison and C.W. Gruber of the University of Cincinnati, and to Dr. R. Lee Byers of the Exxon Research and Engineering Co. He is also grateful for several classes of students on whom these ideas could be tested. And to his wife, Helen, for insuring the presence of an atmosphere of tranquillity which is so essential to work of this kind.

William Licht

CONTENTS

AIR POLLUTION CONTROL ENGINEERING

CHAPTER 1

PARTICULATE COLLECTION IN AIR POLLUTION CONTROL

I. AIR QUALITY STANDARDS

The Clean Air amendments of 1970 to the Clean Air Act of 1963 of the U.S. Congress provide that the Administrator shall promulgate national primary and secondary ambient air quality standards for each type of pollutant. The Act defines: "National primary ambient air quality standards shall be ambient air quality standards the attainment and maintenance of which...are requisite to protect the public health." And "National secondary ambient air quality standards shall specify a level of air quality the attainment and maintenance of which ...is requisite to protect the public welfare from any known or anticipated adverse effects associated with the presence of such air pollutant in the ambient air." Protection of the public welfare has been taken to mean protection against effects on soil, water, vegetation, materials, animals, weather, visibility, and personal comfort and well-being.

Pursuant to this requirement, the Administrator of the Environmental Protection Agency (EPA) on April 30, 1971 promulgated the following ambient air quality standards for particulate matter

Primary standard:
- 75 micrograms per cubic meter, annual geometric mean
- 260 micrograms per cubic meter, as a maximum 24-hour concentration not to be exceeded more than once a year.

Secondary standard:
- 60 micrograms per cubic meter, annual geometric mean
- 150 micrograms per cubic meter, as a maximum 24-hour concentration not to be exceeded more than once a year.

The Clean Air Amendments of 1970 further require that each state must devise and establish an implementation plan by which it intends to attain the ambient air quality standards, not only for particulate matter but for all kinds of pollutants. A state is free to adopt more strigent ambient air quality standards than those promulgated as national standards, but may not adopt less stringent standards.

Provision is made in the 1970 Amendments for revision in the national ambient air quality standards whenever, in the judgment of the Administrator, there is sufficient evidence to warrant a revision.

II. EMISSION REGULATIONS

The only practical way to insure attainment of the legally specified ambient air quality standards is to place a suitable limitation upon the amount of each pollutant which may be emitted from each source where it arises. Thus, a necessary part of any air pollution control implementation plan is a set of emission regulations which may be applied to individual sources. The 1970 Amendments provide that the Administrator shall promulgate Federal standards of performance for new stationary sources, and for moving sources. States may, again, adopt emission regulations more stringent, but not less stringent, than those set forth by the Federal standards of performance.

The relationship between the individual emissions of a given pollutant from the multiplicity of sources within an air quality region, and the resulting average concentration of that pollutant in the ambient air is obviously a complex one. It must involve a reasonably accurate and complete listing of all the sources (their location, height of stack, rate of emission etc.), known as an "emission inventory", as well as meteorological date covering the direction and velocity of winds, presence of inversion, etc. Our knowledge of this overall relationship is incomplete. Various dispersion models are being developed and tested with computer calculations.

Until a reasonably accurate quantitative relationship is established between the strength of individual point source emissions and overall ambient air quality, it is difficult to determine appropriate standards for emission regulations. Hence the 1970 Amendments state that

> The term 'standard of performance' means a standard for emissions of air pollutants which reflects the degree of emission limitation achievable through the application of the best system of emission reduction which (taking into account the cost of achieving such reduction) the Administrator determines has been adequately demonstrated.

Acting under these provisions of the law, the Administrator promulgated the first set of Standards of Performance for New Stationary Sources (NSPS), on December 23, 1971 relative to particulate matter, as follows

A. Fossil fuel-fired steam generators

No owner or operator...shall cause the discharge into the atmosphere of particulate matter which is:

(a) In excess of 0.10 lb per million B.t.u. heat input (0.18 g. per million cal.) maximum 2-hour average.

(b) Greater than 20 percent opacity, except that 40 percent opacity shall be permissible for not more than 2 minutes in any hour. (Except where the presence of uncombined water is the only reason for failure to meet these requirements).

B. Incinerators

No owner or operator...shall cause the discharge into the atmosphere of particulate matter which is in excess of 0.08 gr/s.c.f. (0.18 g/NM^3)corrected to 12 percent CO_2, maximum 2-hour average.

C. Portland Cement Plants

(A) No owner of operator...shall cause the discharge into the atmosphere of particulate matter from the <u>kiln</u> which is:
 (1) In excess of 0.30 lb/ton of feed to the kiln (0.15 Kg/metric ton), maximum 2-hour average.
 (2) Greater than 10 percent opacity (except where the presence of uncombined water vapor is the only reason for failure to meet this requirement).

(B) No owner or operator...shall cause the discharge into the atmosphere of particulate matter from the <u>clinker cooler</u> which is:
 (1) In excess of 0.10 lb/ton of feed to the kiln (0.050 Kg/metric ton), maximum 2-hour average.
 (2) 10 percent opacity or greater.

(C) No owner or operator...shall cause the discharge into the atmosphere of particulate matter from any affected facility other than the kiln and clinker cooler which is 10 percent opacity or greater.*

Subsequently, Federal standards of performance regarding particulate matter have been listed for asphalt concrete plants; catalyst regenerators in petroleum refineries; furnaces in secondary lead smelters, secondary brass and bronze smelters, ferroalloy production; electric arc furnaces and basic oxygen furnaces in iron and steel mills; sludge incinerators for sewage treatment plants; roasters and sintering machines in primary copper and primary zinc smelters; and air tables and thermal dryers in coal cleaning plants. Other sources will continue to be added to the listing as time goes on.

The above regulations are quoted as examples of the type of compliance which is required for new sources. In addition states and cities will have emission regulations for existing sources which

*Federal Register, Dec. 23, 1971.

may differ in form of statement and in stringency from the Federal standards. No doubt some will be revised as a better knowledge is obtained of how emission is related to ambient conditions.

Obviously, it is of the greatest importance that the designer of any particulate collection system or control device learn the emission regulations which apply to the location where the control is to be used. He would be well-advised to design not only to be well within the present regulations but also to anticipate that an even higher degree of collection may be necessary in the future to meet more stringent standards of performance for emissions. He must also become thoroughly familiar with the methods of stack sampling and emission measurement which are to be used in testing whether his equipment is functioning within compliance of the regulations. Some standard test method is usually specified by the appropriate authority, along with the emission regulations.

The air quality standards and the emission regulations are at present always stated in terms of total mass of particulate being emitted per unit time, and usually also in terms of the opacity of the emission stream. No cognizance is taken of the fact that the mass will usually be composed of particles of a wide range of sizes, and perhaps of different composition as well. The coarser particles in an aerosol account for the larger proportion of its total mass. Hence a collection system which is designed to do a good job on the removal of coarser particles, can usually be made to meet the regulations.

However, very fine particles, i.e. of the order of 0.1 μm in size, are those which cause the greater health hazard. Collection of such small particles is a more difficult problem, and one to which increasing attention is being given. It may be that in time emission regulations will be expressed in terms of particle sizes. The design of more sophisticated collectors will be required. This can only be achieved on the basis of a thorough understanding of certain fundamental principles, which are presented in this book.

III. SELECTION OF COLLECTION SYSTEM

There are many kinds of particulate collection devices available on the market. They utilize a number of different principles for accomplishing removal of particles from a gas stream. The differences result mainly from the nature of the force which is applied to a particle in order to collect it. The possible number of fundamentally different forces is not great. Either singly or in certain combinations they constitute the basic mechanisms by which collection is achieved. But these mechanisms are not all successful to the same degree on particles of all sizes.

The selection of appropriate type of collector for a given particulate collection duty must be based upon a knowledge of collector performances which may be anticipated (or predicted) under various operating conditions. There are advantages and disadvantages accompanying the utilization of any particular type of collector embodying certain collecting forces or mechanisms. With regard to performance in terms of collection effectiveness, energy requirements, initial cost, operating cost, ease of installation and maintenance, etc. each of the collectors available finds its most appropriate type of application.

Table 1 is a comprehensive check list of items which must be considered in the design of a system to meet the needs of any air pollution control problem, gaseous or particulate in nature. Obviously a great deal of practical as well as theoretical information will be needed to determine the optimum design of control system for each problem.

It is not the purpose here to present the descriptive details of hardware available. It is assumed that the reader possesses the minimum knowledge of this kind which is needed to understand the application of theoretical principles. There are a number of excellent references which may be consulted for a general survey of types of particulate collectors and their typical performance, costs and other features. Some of these are listed in Table 2.

Table 1

CHECK LIST OF CONSIDERATIONS IN THE DESIGN OF AN AIR POLLUTION CONTROL SYSTEM

I. Survey of the Problem (Existing; Potential)

A. Type of Harmful Effect (Anticipated or Present)

1. Illness and health effects on humans
2. Toxicity to plants and/or animals
3. Damage to material
    a. Corrosion, deterioration, etc.
    b. Soiling
4. Reduction of visibility
5. Odor
6. Annoyances
7. Nature of complaints received

B. Properties of the gas

1. Pollutants present
2. Composition (analysis)
3. Temperature
4. Pressure
5. Humidity and Dewpoint
6. Chemical equilibria among components

C. Flow Rate

1. Quantity of gas to be treated
2. Steadiness of flow; possible fluctuations and start-up, shut-down

D. Particulates

1. Liquid or solids
2. Size distribution of particles
3. Amount of loading of particulates, mass/volume of gas
4. Physical and chemical properties
    a. Composition
    b. Density
    c. Shape
    d. Corrosive
    e. Abrasiveness
    f. Electrical properties
    g. Hygroscopicity
    h. Agglomeration
    i. Adsorption
    j. Combustibility

Table 1 (continued)

- E. Location of plant
    1. Meteorology factors of area
    2. Plant and adjacent property, terrain
    3. Stacks
        a. Height
        b. Location
    4. Existing plant vs. Site Selection for new plant
- F. Degree of Pollution Control Required
    1. Local Standards and Regulations
    2. Tolerable Levels
    3. Future requirements of control system
    4. Methods of testing and measurement
    5. Public image of plant and company
- G. "Excursions" of conditions
    1. Known intermittent operations
    2. Accidental release

II. Solving the Problem

- A. Process Changes
- B. Alternate Materials in process
- C. Selection of Methods of Control of Emissions
    1. Gaseous removal
    2. Particulate removal
    3. Wet or dry system
    4. Single system or combination
    5. Efficiency
- D. Existing control system (if any) and possible adaptation
- E. Use of Pilot Units
- F. Consideration of other pollution which may occur in consequence

III. Design of system

- A. Size of Equipment
- B. Space needed for control unit and auxiliary equipment (fans, ducts, disposal, pumps, power supply etc.)

Table 1 (continued)

C. Materials of Construction

Corrosion allowance

D. Power requirements and pressure drop

E. Controls needed

1. Temperature
2. Humidity
3. Flows
4. Other

F. Maintenance needs

1. Ease of maintenance
2. Time required for maintenance - schedule

G. Flexibility of System - Future needs

H. Safety aspects

I. Availability of standard equipment, or need for special design

J. Cost Estimates

K. Optimizing Design

L. Alternatives

M. Installation permit or "variance"

IV. Performance

A. Testing of installed equipment

B. Operating permit or "variance"

C. Monitoring regular operation

Table 2

REFERENCES ON PARTICULATE COLLECTION

The following references deal especially with the practical aspects, such as descriptions of typical equipment, considerations in selecting equipment, typical performance data and cost data.

1. W.Strauss, *Industrial Gas Cleaning*, 2nd ed., Pergamon Press, Oxford (1975).

   Detailed presentation of theories and equipment for all kinds of gas cleaning.

2. A.Stern, ed., *Air Pollution - Vol. IV*, 3rd ed., Academic Press, New York, (1977).

   Deals generally with control concepts of all kinds, and especially of problems in specific inductries.

3. J.M. Marchello and J.J. Kelly, ed., *Gas Cleaning for Air Quality Control*, Marcel Dekker, Inc., New York (1975).

   General survey of methods, equipment, and application.

4. *"Air Pollution Engineering Manual"* 2nd Edition, Chap. 4 -- J.A. Danielson, Editor. Publication AP-40 Environmental Protection Agency, Research Triangle Park, N.C. (1973).

5. L.S. Chaput, *J. Air Poll. Cont. Assoc.*, 26: 1055 (1976). "Federal standards of performance for new stationary sources of air pollution - a summary of regulations."

6. *"Dust Collection Equipment"* -- G.D. Sargent, Chem. Eng., pp. 130-150, Jan. 27, 1969.

   Excellent survey of types of equipment, performance, sampling methods, costs.

7. *"Designing Air Pollution Control Systems"* -- G.J. Celenza, Ch.E. Progress, 66, no. 11, pp. 31-40, Nov. 1970.

   Good general introduction to equipment and its selection.

8. *"Selection of Gas Cleaning Equipment"* -- C.J. Stairmand, Filtration and Separation, 7, pp. 53-63, Jan./Feb. 1970.

   Good general discussion from British practice.

Table 2. (continued)

9. *"Removal of Grit, Dust, and Fumes from Exhaust Gases from Chemical Engineering Processes"*, Institution of Chem. Eng., pp. CE 310-CE 326, Dec. 1965.

   Much more detailed but somewhat out-of-date version of no. 8

10. *"Estimating the Costs of Gas Cleaning Plants"* -- J.R.F. Alonso, Chem. Eng., pp. 86-96, Dec. 13, 1971.

    Detailed data for cost estimation.

11. *"Capital and Operating Costs of Selected Air Pollution Control Systems"* -- R.B. Neveril, J.V. Price, and K.L. Engdahl, J. Air Poll. Control Assn., 28: 829,963,1069, (1978).

    New method of approach to cost estimation.

CHAPTER 2

CHARACTERIZATION OF PARTICLES AND AEROSOLS

No discussion of the collection of aerosol particles can be undertaken without adequate means available to describe the relevant properties of individual particles, and of the group of particles constituting an aerosol. The most important properties of an individual particle are its size, shape and density. All particles having a specified size, shape and density constitute a "grade" of particles, to be designated as the ith grade. Other related properties of interest are the surface area, volume, mass and chemical composition of a particle.

The assembly of particles making up an aerosol must be described in terms of its overall concentration (e.g. total mass/volume or total number/volume) and of the distribution of grades by number-, volume-, or mass-fraction within each grade.

The techniques by which individual particles are examined, and by which a size- (or more accurately a grade-) distribution analysis is performed, constitute an important and difficult branch of physical measurements. For a detailed discussion of methods, the reader is referred to sources listed in the references [1,2,3,4]. This chapter is devoted to the interpretation of the results of such measurements and their application to the problems of particulate collection.

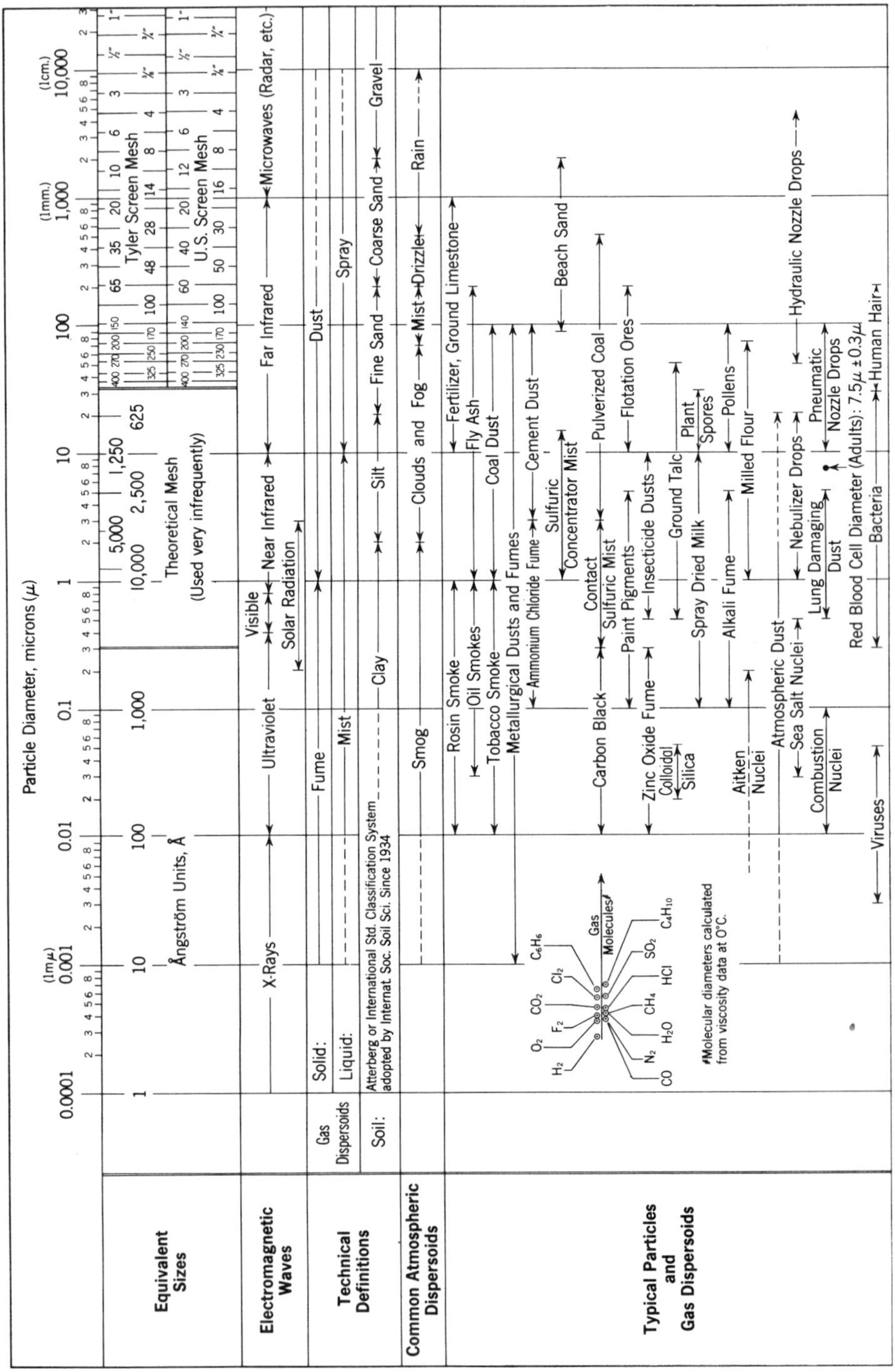
Particle Diameter, microns (μ)
0.0001
0.001 (1mμ)
0.01
0.1
1
10
100
1,000 (1mm.)
10,000 (1cm.)
Equivalent Sizes
Ångström Units, Å
Theoretical Mesh (Used very infrequently)
Tyler Screen Mesh
U.S. Screen Mesh
Electromagnetic Waves
X-Rays
Ultraviolet
Visible
Near Infrared
Far Infrared
Microwaves (Radar, etc.)
Solar Radiation
Technical Definitions
Gas Dispersoids
Solid:
Fume
Dust
Liquid:
Mist
Spray
Soil:
Atterberg or International Std. Classification System adopted by Internat. Soc. Soil Sci. Since 1934
Clay
Silt
Fine Sand
Coarse Sand
Gravel
Common Atmospheric Dispersoids
Smog
Clouds and Fog
Mist
Drizzle
Rain
Typical Particles and Gas Dispersoids
Gas Molecules*
*Molecular diameters calculated from viscosity data at 0°C.
Rosin Smoke
Oil Smokes
Tobacco Smoke
Metallurgical Dusts and Fumes
Ammonium Chloride Fume
Fertilizer, Ground Limestone
Fly Ash
Coal Dust
Cement Dust
Sulfuric Concentrator Mist
Beach Sand
Carbon Black
Contact Sulfuric Mist
Pulverized Coal
Zinc Oxide Fume
Paint Pigments
Insecticide Dusts
Flotation Ores
Colloidal Silica
Ground Talc
Spray Dried Milk
Plant Spores
Alkali Fume
Pollens
Milled Flour
Aitken Nuclei
Atmospheric Dust
Sea Salt Nuclei
Nebulizer Drops
Hydraulic Nozzle Drops
Combustion Nuclei
Lung Damaging Dust
Pneumatic Nozzle Drops
Red Blood Cell Diameter (Adults): 7.5μ ± 0.3μ
Viruses
Bacteria
Human Hair

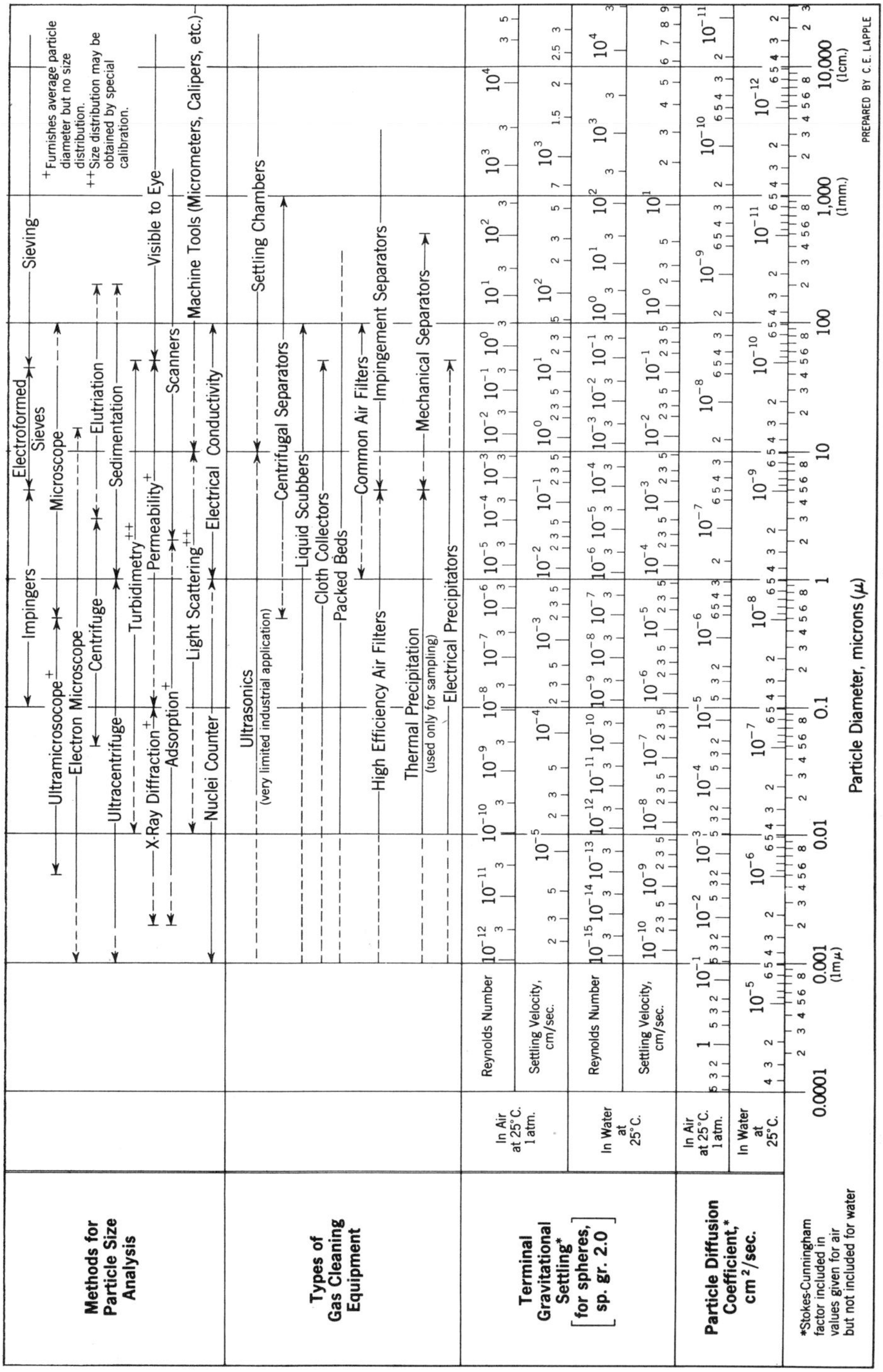

FIG. 1. Characteristics of small particles. Reproduced by permission of Stanford Research Institute.

I. DESCRIPTION OF INDIVIDUAL PARTICLES

The basic tool for examining particles is microscopy, either of the optical or electronic variety. It reveals that solid particles making up most dusts and fumes are usually very irregular in shape, but occasionally correspond to a regular crystalline state. Liquid particles present in mists and sprays tend to be spherical, although they may be more or less distorted in larger sizes. A concise summary of the range of sizes and shapes found for many materials is given in Figure 1. An excellent descriptive source is "The Particle Atlas" by McCrone [5] which is a comprehensive collection of photomicrographs of particles. This can be of great assistance in the identification of the nature and source of particles of unknown origin.

The irregular shapes observed tend to fall into one of three general classes:

(a) Isometric - in which all three overall dimensions of the particle are roughly of the same magnitude;

(b) Flat - in which there are much greater lengths in two dimensions than in the third, e.g. platelets, scales, leaves, etc.

(c) Needlelike - in which there is a much greater length in one dimension than in the other two, as for instance in prisims or fibers.

Examination of a particle under the microscope only reveals the two-dimensional outline which happened to fall at random into the plane of observation. The thickness of the particle may be inferred indirectly. Other methods of observation measure the total volume, total mass, or total surface of a particle, or its mechanical behavior (motion) under the application of certain forces (gravitational, centrifugal, electrostatic.

It is almost impossible to assign a set of numbers to give a complete description of the size and shape of an irregualr particle, indeed it is almost impossible to define these terms. Instead an indirect method of description is usually adopted which is related to the purpose for which the "size" of the particle is to be determined. For example, the rate of settling in air under gravity may be measured for some irregular particles. Then a "size" number may be assigned to each particle on the basis of calculation of the diameter of a spherical particle of the same density which would settle at the same speed as that measured. For the purpose of discussing the movement of such a particle under the influence of gravity, the diameter so calculated may be referred to as the "effective size" of the particle. This kind of calculation is explained in Chapter 4.

Figure 1 summarizes the types of size-analysis techniques and gives an indication of the range of conditions under which each is applicable. It is noted that in most cases some property is being measured which is rather indirectly related to the "size" of the particle. A numerical value for the latter is obtained on the basis of some sort of "equivalent behavior".

It is convenient to try to describe a real irregular particle in terms of the size of some simple geometrical shape which will behave in the equivalent manner. The simple geometrical shapes employed are:

Sphere:	diameter = d
Cube:	side = d
Cylinder:	length = ℓ, diameter = d
Ellipsoid of revolution:	polar radius = p, equatorial radius = r

$$\beta = \frac{p}{r} \qquad \begin{array}{ll} \text{Prolate} & \beta > 1 \\ \text{Oblate} & \beta < 1 \\ \text{Sphere} & \beta = 1 \end{array}$$

An isometric particle may be approximated by either a sphere or cube, a disk-like particle by either a cylinder with ℓ/d very small or an oblate ellipsoid with $\beta \rightarrow 0$, and a needlelike particle by a cylinder with large ℓ/d or prolate ellipsoid with $\beta \rightarrow \infty$. The system of ellipsoids may be used to cover almost all cases, but the geometrical calculations for them tend to become rather complicated.

With these concepts in mind, the following definitions of particle "size," in common use, may be stated:

> "Projected diameter" - the diameter of a circle having the same area as the image of the particle projected parallel to the plane of the microscope view.
>
> "Equivalent diameter" - the diameter of a sphere having a volume equal to that of the particle.
>
> "Sedimentation diameter" - the diameter of a sphere of equal density, having the same settling velocity as the particle in a specified fluid.
>
> "Aerodynamic diameter" - the diameter of a sphere of unit density (1 gm/cm) having the same settling velocity in air as the particle.

The calculation of aerodynamic diameter is discussed in Chapter 4.

Other "diameters" are defined with reference to the particle image under the microscope as indicated by the sketch:

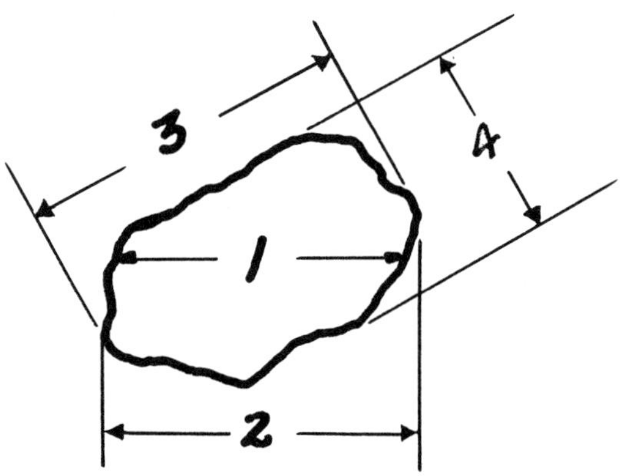

(1) Martin's diameter, divides projected area into equal parts along line parallel to base of field of view.

(2) Feret's diameter, must be longest dimension along line parallel to base of field of view.

(3) Longest diameter, without regard to direction.

(4) Shortest diameter, without regard to direction

The "sphericity" of a particle is defined as: ϕ_s = surface area of a sphere having the same volume as the particle ÷ surface area of the particle. The value of ϕ_s is always less than one. For a cube $\phi_s = 0.806$. For a cylinder $\phi_s = 2.62\ (\ell/d)^{2/3}\ /(1 + 2\ \ell \div d)$ and for $\ell/d = 10$, $\phi_s = 0.579$. Some experimentally measured values for ϕ_s are given in Table 1.

Table 1

VALUES OF SPHERICITY [6]

Material	ϕ_s
Sand	0.600, 0.681
Iron catalyst	0.578
Bituminous coal	0.625
Celite cylinders	0.861
Broken solids	0.63
Sand	0.534-0.628
Silica	0.554-0.628
Pulverized coal	0.696

The surface, volume and mass of a particle may be related to an appropriately defined diameter d_p, such that

$$\text{surface} \quad \alpha\ d_p^2$$

$$\text{volume} \quad \alpha\ d_p^3$$

$$\text{mass} \quad \alpha\ \rho_p\ d_p^3$$

The "specific surface" is usually defined as surface per unit volume (occasionally per unit mass) and is given as $\Omega \propto 1/d_p$. Values are:

$$\Omega = 6/d_p \text{ for sphere } (d_p = \text{diameter}), \text{ or cube } (d_p = \text{side})$$

$$\Omega = \frac{2}{\ell}\left(1 + \frac{2\ell}{d}\right) \text{ for cylinder}$$

For some purposes e.g. catalysis, adsorption, etc., the question of what constitutes the "surface" of a particle requires closer study. These phenomena occur at a size scale of the order of magnitude of atomic and molecular dimensions. The presence of surface imperfections, active sites, and other situations at this size scale may make the effective surface quite different from and far greater than the gross geometric surface which is related to drag forces, light scattering, aerodynamic capture, and other phenomena. These are characterized by a surface interaction on the size scale of the particle dimension itself which is usually at least several orders of magnitude larger than atomic or molecular dimensions.

Particles frequently have a tendency to agglomerate, with the resulting clump being "the particle" whose behavior is to be studied or controlled. The presence of clumps can usually be detected by microscopic examination. Some sort of appropriate dimension can be assigned to the clump acting as a single particle. Care must be taken in using such size values, for the apparent density of the agglomerate will be much less than that of the material of which the particles become individuals again in the stream flowing through the collecting equipment.

The chemical composition of the material of the particle determines its density as well as a number of other properties of importance in the collection process. Among these are its hardness (tendency to absorb moisture, become sticky), and electrical resistivity (tendency to acquire and hold electrical charges), all of which have consequences which may be of great practical importance in the collection process.

II. SIZE DISTRIBUTION IN AEROSOLS

In order to make calculations such as will be discussed in Chapter 3 for the collection of any given aerosol, it is necessary to know the grade-distribution (or "size distribution") expressed say in terms of mass-fractions m_i:

$$m_i = f_i\,(d_{p_i}) \qquad \text{where } \Sigma\, m_i = 1.$$

Experimental methods are needed for determining the mass or number of particles of each size grade in a representative sample of the aerosol. Such methods are described in detail in references [1,2,3,4 and 7]. The size values for d_{p_i} will usually be expressed in terms of some simple geometrical shape having "equivalent behavior" in the sense of the measuring technique employed.

For the purpose of calculating the total emission by weight from a collector, an accurate knowledge of the grade distribution of the coarser particles (e.g., larger than 2 μm) is most important, because these account for the overwhelming proportion of the total weight of many aerosols. However, increasing recognition is being given to the fact that most of the bad health effects on humans and animals arise from the inhaling of very fine particles (e.g., less than 1 μm). Greater attention must be given to keeping these out of the ambient air, hence greater importance will be attached to more accurate measurement of grade-distribution in the very fine size range. The problem has been concisely summarized by Iammartino [8].

It is frequently handy to characterize an entire aerosol by some kind of "average" or "mean" particle size. There are many ways listed below in which an average or mean value may be defined. One must be careful to use a mean value which correctly relates to the behavior of the aerosol under the particular circumstances being considered.

A. Raw Data For Grades

An examination of an aerosol sample by almost any of the usual techniques will give size-distribution information in terms of size intervals or ranges. That is raw data will be in the form of the number of particles, or the total mass of particles, falling within a specified range of sizes. Alternatively the total number (or mass) of particles smaller than (or larger than) a specified size may be given. Very rarely (as when counting glass beads under the microscope) will it be possible to identify individual particles by exact sizes, and to tabulate the number of particles found to be of each precise size. Data of the first kind may be displayed graphically as a histogram, in which the intervals of size-range need not all be equal. Figure 2 is a typical illustration, the raw data for which are tabulated in Example 1.

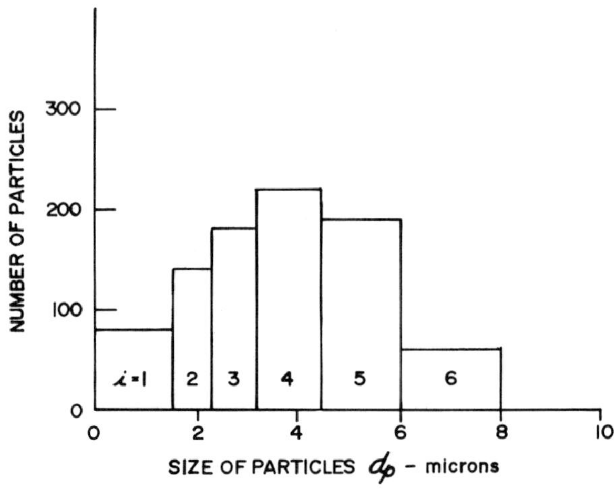

FIG. 2. Histogram of Particle Count - Example 1.

In general, it is reasonable to expect the size-distribution to indicate a modal (most frequently occuring) value, or perhaps occasionally a bi-modal form. If the raw data do not clearly indicate any mode, the data may be suspect as inadequate or inaccurate and further checking of them is in order. It is also reasonable to expect that there will be relatively few very small particles, and relatively few very large particles, so that the mode is not likely to be at either extreme end of the total size range.

B. Frequency and Cumulative Fractions

The number, or weight, values displayed by the raw data may be transformed into (a) fractional, or frequency, values, or (b) cumulative values. This will be developed first in terms of number data, and the corresponding calculations for weight data presented as an example below.

Let n_i = number of particles observed in i^{th} interval

Σn_i = total number of particles of all sizes observed

Then: (a) Fractional or frequency values:

$f_i = \frac{n_i}{\Sigma n_i}$ = fraction of particles in i^{th} interval

where $\Sigma f_i = 1$. Thus, f_i also represents the frequency of occurrence of particles in the ith interval. (1)

Or: (b) Cumulative values:

$$F_j = \frac{\sum_{i=1}^{i=j} n_i}{\sum_{i=1}^{n} n_i} \qquad F_j = \sum_{i=1}^{j} f_i \qquad \text{and } F_n = 1 \tag{2}$$

F_j = fraction of all particles which are smaller than the largest size in the jth interval, called the "cumulative-less-than" fraction. Similarly, a "cumulative-greater-than" fraction may be defined. The raw data, especially in terms of weights, may sometimes be obtained in the cumulative form also.

The raw data of the particle count may be recalculated into frequency values and plotted against the particle size representing the mid-point of each interval. This is shown on Figure 3. There is then a temptation to draw a smoothed curve through these points. This temptation should be resisted, because such a curve would be misleading. It would imply that at any point in the total range of particle sizes one may take a size interval to be as small as desired, and that for this interval around the selected point, the curve would give a correct value for the frequency. Obviously this is not the case, for the smaller the interval chosen, the smaller the value of f must be. The only meaning that could be given to a smoothed curve drawn through the histogram points would be that the ordinates could represent the frequency of occurrence of particles of the same size interval as measured, but around other midpoints as might be read on the the curves. But if the original raw data were not determined for size intervals of equal width, then this would not apply.

Two important questions then arise:

(a) How can we determine the frequency of occurrence of particles lying within any size-interval other than those actually counted?

(b) In the limiting case, can we determine the frequency of occurrency of particles of a specific size?

These questions are best explored through the use of the cumulative-less-than function F and its graph.

The raw data are used to calculate the F-values, as defined above, and these are each plotted against the particle size value which is the upper limit of each size-interval counted. This is shown in Figure 4. Here a smooth curve may be drawn without question and its meaning is clear. If we take any two points, designated as (a) and (b), upon this curve we define an interval of particle sizes

d_{P_a} to d_{P_b}. The difference between the F values ($F_a - F_b$) is clearly equal to the value of f for this interval. The calculation may also be represented in terms of the slope of the F-curve thus:

$$f_{a-b} = F_a - F_b = \int_{F_b}^{F_a} dF = \int_{d_{P_b}}^{d_{P_a}} \frac{dF}{dd_p} \times dd_p = \int_{d_{P_b}}^{d_{P_a}} p\ dd_p \tag{3}$$

The F-curve also makes it easy to determine the number median particle size (NMD), which is the value of d_p for which F = 0.50. Half of the particles are larger and half smaller than this size.

C. Density Functions

1. Population Density

Equation (3) implies the definition of a useful function $p = \frac{dF}{dd_p}$, called "population density," having dimensions of (size^{-1}). It is a continuous function of particle-size, and from the definitions it is evident that:

$$F = \int_0^{d_p} p.dd_p \quad \text{and} \quad \int_0^{\infty} p.dd_p = 1 \tag{4}$$

$$\frac{dp}{dd_p} = \frac{d^2F}{dd_p^2} \tag{5}$$

In the limits as $d_p \to o$, or $d_p \to \infty$, f must approach zero, hence $p \to o$ and the F-curve must be S-shaped with a point of inflection. This point of inflection occurs at a particle size for which p is a maximum i.e., where

$$\frac{d^2F}{dd_p^2} = \frac{dp}{dd_p} = 0 \tag{6}$$

The p-function may be constructed by measuring the slope of the F-curve, or approximately by calculating the ratio $(F_a - F_b)/(d_{p_a} - d_{p_b})$ to represent the slope at the point $(d_{p_a} + d_{p_b})/2$. A graph of p against particle size will not be unlike that of the histogram of f_i vs. d_p in general appearance, but the p vs. d_p plot may be drawn validly as a smooth curve. The area under this curve between two particle sizes gives the value of f for that size interval. Figure 3 includes a curve for p as calculated from Fig. 4. Example 1 gives all the data and calculations for Figs. 2,3 and 4.

a. Example 1

Grade Designation-i	Range of Sizes-μm	Number of Particles-n_i
1	0-1.5	80
2	1.5-2.3	140
3	2.3-3.2	180
4	3.2-4.5	220
5	4.5-6.0	190
6	6.0-8.0	60
	> 8.0	0
	Total	870

The calculated values for the frequency plot Fig. 3 and the cumulative plot Fig. 4 are obtained as follows:

Grade Designa-tion-i	Midpoint of Size Range - d_p - μm	Number Fraction-f_i (Eqn. 1)	Upper Bound size of range-μm	Cumulative Fraction less than - F_j (Eqn. 2)
1	0.75	80/870=0.092	1.5	80/870=0.092
2	1.90	140/870=0.161	2.3	0.092+0.161=0.253
3	2.75	180/870=0.207	3.2	0.253+0.207=0.460
4	3.85	220/870=0.253	4.5	0.460+0.253=0.713
5	5.25	190/870=0.218	6.0	0.713+0.218=0.931
6	7.00	60/870=0.069	8.0	0.93 +0.07 =1.000
		Total 1.00		

The calculated values for the population density plot, also shown on Fig. 3 are as follows. The p-values are plotted against

d_p for the midpoint of each interval.

Grade Designa-tion-i	F-interval ΔF	d_p-range Δd_p - m	$\frac{\Delta F}{\Delta d_p} \simeq p.- \mu m^{-1}$ (Eqn. 5)
6	1-0.931 = 0.069	8.0-6.0 = 2.0	0.035
5	0.931-0.713=0.218	6.0-4.5 = 1.5	0.145
4	0.713-0.460=0.253	4.5-3.2 = 1.3	0.195
3	0.460-0.253=0.207	3.2-2.3 = 0.9	0.230
2	0.253-0.092=0.161	2.3-1.5 = 0.8	0.201
1	0.092-0 =0092	1.5-0 = 1.5	0.061

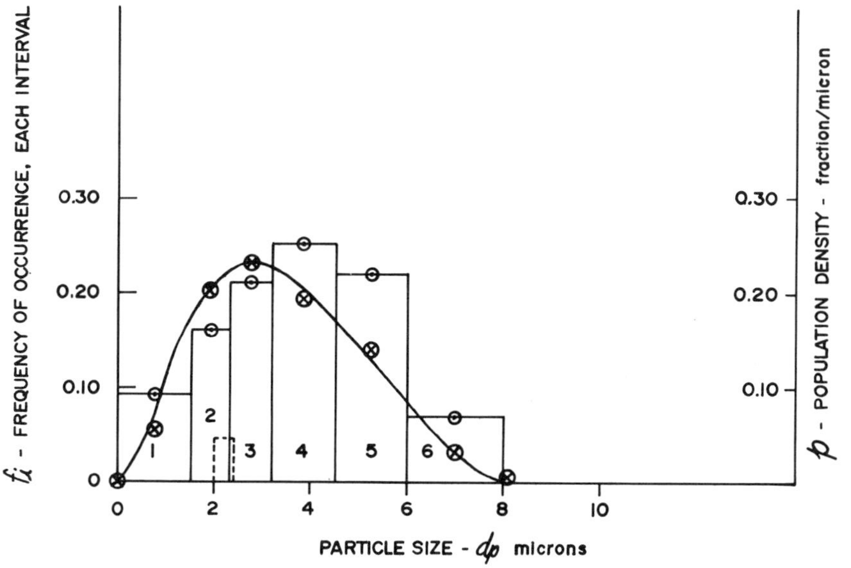

FIG. 3. Relation between frequency of interval occurance and population density, Example 1. Bar chart: calculated discrete values of f_i, curve: population density - p.

Note that p was estimated by the approximate method of calculating the slope as $\Delta F/\Delta d_p$. For convenience, the same intervals and F values were used as above, but, of course, other intervals and F values read from the curve (Fig.4) could have been used just as well. The smoothness of the p-curve which can be drawn with such points

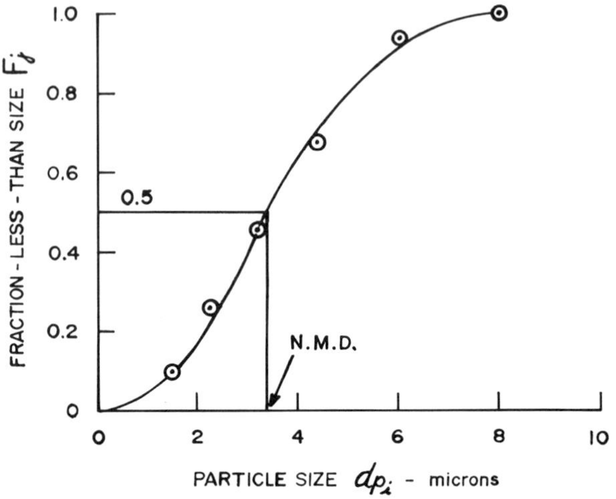

FIG 4. Cumulative fraction - less-than particle size, Example 1. N.M.D. = Number median diameter = 3.3 microns

is one test of the satisfactory use of the approximate slope calculation.

As an example of the use of the p-plot, determine the frequency of occurrence in the particle size-interval between 2.0 and 2.4 microns. This involves evaluation of the area under the p-curve:

$$f_{2.2-2.4} = \int_{2.0}^{2.4} p.dd_p \approx \frac{0.205 + 0.225}{2} (2.4-2.0) = 0.043 \tag{6}$$

The value of f is shown as the dotted bar on the histogram of Fig. 3. The integral is estimated by using the area under the p-curve on Fig. 3, using the average value of p (0.215) over the interval as ordinate.

It should now be clear that the question of what is the frequency of occurrence of particles of a precise size cannot be answered from raw data of the kind we are discussing. We may take the interval $(d_{p_a} - d_{p_b})$ to be as small as we like (e.g. $df = pdd_p = dF$) and calculate an f for it, but to take it as zero will render the calculation impossible. Unless the raw data have consisted of an actual count of individual particles by size, we cannot say what such an f-value is. It will be seen in what follows that the data are almost always handled best in terms of the F and p functions, or their counterparts in terms of particle mass instead of numbers.

The real importance of the p-function lies in the fact that it has been observed to follow consistently certain rather simple and well-defined forms which are characteristic of types of aerosols. The example above is typical in that the p-curve looks something like a skewed normal probability function. A few basic types of equations have been found to give accurate representations of p, hence also of F, or vice versa. These are discussed below. With the aid of such equations various kinds of mean and median particle sizes may be defined and readily calculated. The calculation of collector performance upon an aerosol described by a p-function is also facilitated.

2. Weight Density

Grade distribution data given in terms of numbers of particles may be converted into terms of weight of particles (or vice versa) by assuming that the weight of a particle is proportional to the cube of its size. Then either all particles in a sample must be taken to have the same density or the actual density of the different kinds of particles present must be determined and used to convert volume to weight. Plots of frequency, cumulative, or weight-density (analogous to population-density) values may then be defined and constructed just as was done for the number data. The definitions are, assuming all particles to have the same density:

$$g_i = \frac{n_i d_{p_i}^3}{\sum_{i=1}^{n} n_i d_{p_i}^3} = \text{weight fraction of particle in } i^{th} \text{ grade} \tag{7}$$

$$G_j = \sum_{i=1}^{j} g_i = \text{cumulative-less-than fraction by weight up to and including } j^{th} \text{ grade} \tag{8}$$

$$q = \frac{dG}{dd_p} = \text{weight-density function} \tag{9}$$

The calculations are illustrated in Example 2.

a. Example 2. Using the raw data of Example 1, convert it into size distributions in terms of weight of particles and display the corresponding graphs of G_j and q. Assume all particles have the same density.

The calculations of g_i and G_j are performed as tabulated:

Grade Designa-tion-i	Midpoint size of range d_{Pi} - μm	$(d_{Pi})^3$ -(μm)^3	$n_i \times d_{Pi}^3$	g_i (Eqn. 7)	G_j (Eqn. 8)
1	0.75	0.422	34	5×10^{-4}	5.2×10^{-4}
2	1.90	6.86	960	0.015	0.016
3	2.75	20.79	3742	0.057	0.073
4	3.85	57.07	12555	0.192	0.265
5	5.25	144.7	27493	0.421	0.686
6	7.00	343	20580	0.315	1.000
		Totals	65364	1.000	

Following the calculation of G_j, the q-values are approximated by $\Delta G/\Delta d_p$ in the same manner as was done for p in the preceding example.

Interval-i	G-interval ΔG	dp-range Δdp-μm	$\Delta G/\Delta dp \approx q$ (μm)$^{-1}$ (Eqn. 9)
6	1 - 0.686 = 0.314	2.0	0.157
5	0.686 - 0265 = 0.421	1.5	0.281
4	0.265 - 0.073 = 0.192	1.3	0.148
3	0.073 - 0.016 = 0.057	0.9	0.063
2	0.016 - 0.00052 = 0.015	0.8	0.019
1	0.00052 - 0 = 5.2×10^{-4}	1.5	3.5×10^{-4}

Figures 5 and 6 repeat the p and F values and include the q and G values for comparison. It is typical that the q and G (mass) curves are displaced toward the right with respect to the p and F (number) curves.

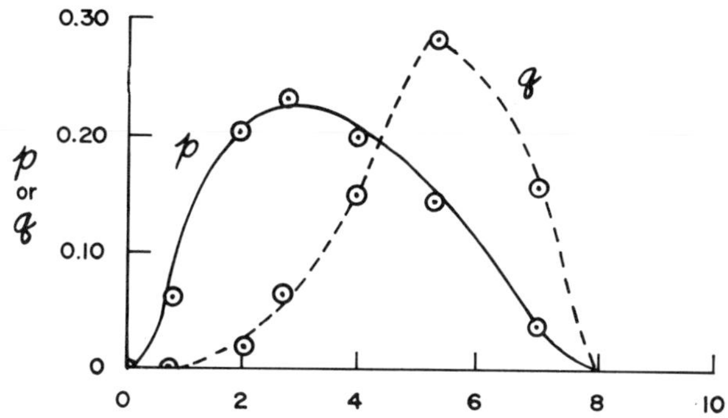

FIG. 5. Density Functions - Examples 1 and 2, p = population density, q = weight (mass) density.

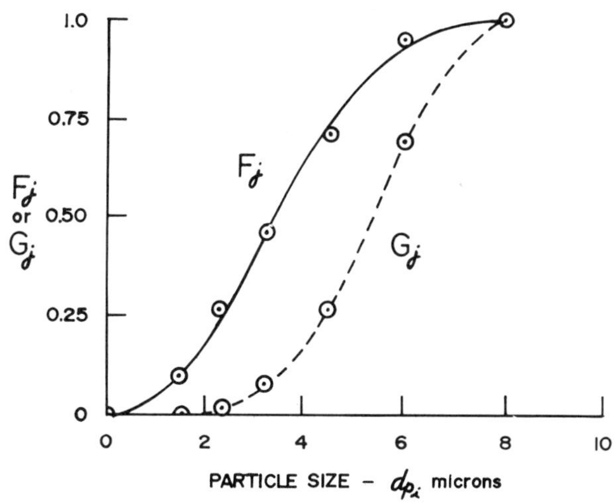

FIG. 6. Relation between number and mass cumulative fraction - less-than size. F_j = number fraction - less-than, G_j = weight fraction - less-than

D. Mode, Median and Mean Values

It is useful to describe an aerosol by means of a single particle-size which represents in some fashion a typical size of the range of particles included in it. This may be done in terms of either a modal, median or mean value.

Where the frequency distribution (f) possesses a single well-defined peak value such as on Fig. 3, this most frequently occuring size is called the mode. For the sample data given the mode is about 2.8 μm. Occasionally there will be two peak values in the frequency distribution, a so-called bimodal pattern. When this occurs, it is significant enough to be emphasized by quoting both modal particle sizes, and to investigate the reasons for their existence. This is discussed further at the end of this chapter.

A median size which divides the sample into two equal portions may be defined in terms of cumulative distributions either by number of particles (F) or by mass of particles (g). Thus: (a) the number median (NMD) particle size is that for which $F = 0.50$, (b) the mass median (MMD) particle size is that for which $G = 0.50$. For the data of Example 1 the number median size is about 3.3 μm, and the mass median size is about 5.3 μm. These values may easily be read from a plot like Fig. 4.

In addition to the mode and median values, a mean value particle size may be used to represent an aerosol. There are a number of different definitions of mean size in common use. The appropriate definition to employ in a given instance will depend usually upon the particular aspect of aerosol behavior which is under consideration. Whitby [9] has set forth a scheme by which any of the definitions of a mean size may be generated as the need arises. This scheme, somewhat modified, is as follows:

(a) Define a moment distribution factor, as the property of an individual particle which is being measured, which is proportional to a certain power of particle size d_p^m:

Size	d_p	1st moment	$m = 1$
Surface	$\alpha\ d_p^2$	2nd moment	$m = 2$
Volume, or Mass	$\alpha\ d_p^3$	3rd moment	$m = 3$

(b) Define a weighting factor, as the total property measured, in all particles of the ith size, which is proportional to $(n_i d_{p_i}{}^w)$.

Total number	n_i	$w = 0$
Total size	$n_i d_{p_i}$	$w = 1$
Total surface	$n_i d_{p_i}^2$	$w = 2$
Total volume, or mass	$n_i d_{p_i}^3$	$w = 3$

where n_i = number of particles of ith size.

(c) Define mean size by selecting a combination of the above factors according to:

$$\text{Mean Size} = \left[\frac{\sum_i (\text{Moment Distribution x Weighting})}{\sum_i (\text{Weighting})}\right]^{1/m}$$

$$\bar{D}_{m,w} = \left[\frac{\sum_i (n_i d_{pi}{}^{m+w})}{\sum_i (f_i d_{p_i}{}^{w})}\right]^{1/m}$$

and using (1):

$$D_{m,w} = \left[\frac{\sum_i (f_i d_{pi}{}^{m+w})}{\sum_i (f_i d_{p_i}{}^{w})}\right]^{1/m} \qquad m \neq 0 \qquad (10)$$

and using (7) also:

$$\bar{D}_{m,w} = \left[\frac{\sum g_i d_{p_i}^{\,m+w-3}}{\sum g_i d_{p_i}^{\,w-3}}\right]^{1/m}$$

Some examples of commonly used means, obtained by specifying m and w are listed below:

"average particle size" w = 0, m = 1

$$D_{1,0} = \frac{\sum n_i d_{p_i}}{\sum n_i} = \sum f_i d_{p_i} \tag{11}$$

"surface mean" w = 0, m = 2

$$\bar{D}_{2,0} = \left[\frac{\sum n_i d_{p_i}^{\,2}}{\sum n_i}\right]^{1/2} = \left[\sum f_i d_{p_i}^{\,2}\right]^{1/2} \tag{12}$$

"volume mean" w = 0, m = 3

$$\bar{D}_{3,0} = \left[\frac{\sum n_i d_{p_i}^{\,3}}{\sum n_i}\right]^{1/3} = \left[\sum f_i d_{p_i}^{\,3}\right]^{1/3} \tag{13}$$

"surface" to diameter mean" w = 1, m = 1

$$D_{1,1} = \frac{\sum n_i d_{p_i}^{\,2}}{\sum n_i d_{p_i}} = \frac{\sum n_i d_{p_i}^{\,2}}{\sum f_i d_{p_i}} \tag{14}$$

Sauter mean diameter: $w = 2,\ m = 1$

"volume/surface"

$$D_{1,2} = \frac{\sum n_i d_{p_i}^3}{\sum n_i d_{p_i}^2} = \frac{\sum f_i d_{p_i}^3}{\sum f_i d_{p_i}^2} = \frac{1}{\sum g_i / d_{p_i}} \tag{15}$$

DeBrouckere diameter: $w = 3,\ m = 1$

"mass mean"

$$\bar{D}_{1,3} = \frac{\sum n_i d_{p_i}^4}{\sum n_i d_{p_i}^3} = \frac{\sum n_i d_{p_i}^4}{\sum f_i d_{p_i}^3} \tag{16}$$

Mugele and Evans [10] have also stated similar schemes for defining mean sizes, but in terms of continuous rather than discrete size distributions. The definitions are, therefore, expressed in the form of integration terms rather than summations. Thus

$$\bar{D}_{m,w} = \left[\frac{\int_0^1 d_p^{\,m+w}\, dF}{\int_0^1 d_p^{\,w}\, dF} \right]^{1/m} \qquad m \neq 0 \tag{17}$$

There are, of course, interrelationships which exist between median values and mean values for a given aerosol. The exact relation depends, however, upon the size distribution which is present. This is discussed below under the topic of size distribution functions.

a. Example 3. Using the data from Examples 1 and 2, calculate NMD, MMD, the "average particle size" $\bar{D}_{1,0}$, the "volume mean" $\bar{D}_{0,3}$, the Sauter mean $\bar{D}_{1,2}$ and estimate the specific surface Ω of the material.

The median diameters may be read directly from Figure 6, as follows: NMD = 3.4 μm (at F = 0.50); MMD = 5.3 μm (at G = 0.50). The calcualtion of average particle size, according to (11):

i	d_{p_i}	n_i	f_i	$n_i d_{p_i}$	$f_i d_{p_i}$
1	0.75	80	0.092	60	0.069
2	1.90	140	0.161	266	0.306
3	2.75	180	0.207	495	0.569
4	3.85	220	0.253	847	0.974
5	5.25	190	0.218	997	1.144
6	7.00	60	0.069	420	0.483
		870	1.00	3085	3.545

$$D_{1,0} = \frac{\sum n_i d_{p_i}}{\sum n_i} = \frac{3085}{870} = 3.55 \ \mu m$$

$$\bar{D}_{1,0} = \sum f_i d_{p_i} = 3.55 \ \mu m$$

The calculation of "volume mean size" according to (13), and the calculation of $\bar{D}_{1,2}$ according to (15), require in addition:

$f_i d_{p_i}^3$	$g_i/d_{p_i}^3$	$f_i d_{p_i}^2$	g_i/d_{p_i}
0.04	0.00118	0.052	0.00067
1.10	0.00219	0.581	0.00789
4.31	0.00274	1.57	0.0207
14.44	0.00336	3.75	0.0499
31.54	0.00291	6.01	0.0802
23.67	0.00092	3.38	0.0450
75.10	0.01330	15.34	0.2044

$$\bar{D}_{0,3} = \left[\sum f_i d_{p_i}{}^3\right]^{1/3} = (75.10)^{1/3} = 4.22\ \mu m$$

or

$$\bar{D}_{0,3} = \left[\frac{1}{\sum g_i/d_{p_i}{}^3}\right]^{3} = \left(\frac{1}{0.0133}\right)^{1/3} = 4.22\ \mu m$$

$$\bar{D}_{1,2} = \frac{\sum f_i d_{p_i}{}^3}{\sum f_i d_{p_i}{}^2} = \frac{75.10}{15.34} = 4.89\ \mu m$$

or

$$\bar{D}_{1,2} = \frac{1}{\sum g_i/d_{p_i}} = \frac{1}{0.2044} = 4.89\ \mu m$$

The specific surface may be estimated from $\bar{D}_{1,2}$ as $\Omega = 6/\bar{D}_{1,2} = 6/4.89 = 1.23\ \mu m^{-1}$ or $1.23 \times 10^4\ cm^2/cm^3$.

III. SIZE DISTRIBUTION FUNCTIONS

A. Recurring Patterns

Where many examples of size distribution data are examined, it becomes evident that typical patterns in the frequency (f, or g) or cumulative (F or G) plots recur rather regularly. A particular pattern of size distribution seems to be associated with the nature of an aerosol. Aerosols naturally occurring in the ambient air seem to follow one pattern, those generated by mechanical disintegration of solid bodies another, those generated by atomization of liquids another, and so on. It is very useful to have a technique for identifying such patterns and expressing then in mathematical form, particularly when that form can be a rather simple one.

It is most fruitful to examine these patterns in terms of the "density functions" p, or q, or the cumulative functions F or G. The sketches of size distribution shown in Fig. 3, 4, 5 and 6 represent typical characteristics of such plots. The distribution (p vs. d_p, Fig. 3) resembles in a general way the bell-shaped plot corresponding to the so-called Gaussian or "normal" distribution. However, instead of being symmetrical about a central particle size, it is usually skewed toward the larger particle sizes. Also, sometimes it may be necessary to take into account that there is a limiting small particle size below which p is zero, and/or that there is a maximum particle size present. The cumulative plots (f vs. d_p, Fig. 4) are typically S-shaped and must be related to the density plots by the Equations (3) and (9).

Whenever possible, it is very convenient to have a simple equation to represent a given distribution pattern. This equation may give either p, or F, (or either q or G) as a function of d_p and two or more constants or coefficients. Ideally the simplest kind of function would involve only two constants. One of these should represent the general size-magnitude of the aerosol, i.e., some kind of a defined mean particle size. The other constant should represent the spread of the size range about this mean. A third constant will, however, usually be needed whenever it is necessary to represent explicitly a limit particle-size (either maximum or minimum) in the aerosol.

Calculations of collector performance based upon grade efficiency curves, as will be explained in Chapter 3, are greatly facilitated by the use of size distribution equations. When the particle-size of the incoming aerosol and the grade efficiency of the collector may each be expressed by simple functions, computer programs may easily be written to calculate the overall efficiency of the collector and to predict the size distribution in the product aerosol emitted.

A number of semi-empirical formulas have been found to possess the features required for an adequate representation of certain types of aerosols. Some of these are listed as follows, with an example of their applicability. (a) Log-normal probability distribution. Two constants. Ambient aerosols; many process dusts. The most commonly used and widely applicable function. (b) Modified "upper limit" log-normal. Three constants. Used when it is necessary to specifically represent maximum particle size, e.g. in sprays. (c) Weibull probability distribution. Three constants. Process dusts, especially those having a limiting small-sized particle. (d) Rosin-Rammler formula. Two constants. Comparatively coarsely dispersed dusts and mists. Mathematically a special case of (c) above. (e) Roller's formula. Two constants. Powdered industrial materials. (f) Nukiyama and Tanasawa's formula. Three constants. Mechanically dispersed mists (atomization).

In order to select one of these to use for representing a given set of data, it is necessary first to have a knowledge of the general features and distinctive characteristics of each one. Then there must be a specific method for testing the goodness of fit and for determining the best values of the constants in each one. This information is presented in the next section.

It should also be pointed out that one of these equations may be useful in representing other kinds of relationships. For example, in Chapter 5 there are certain graphs relating to the aerodynamic capture of particles which have the typical S-shaped appearance of the cumulative plots. These can be represented usefully by one of the formulas just listed.

B. Specific Functions

A size-distribution function may be given initially in one of four ways: population-density, fraction by number of particles: p; weight-density, fraction by mass (or volume) of particles: q;

cumulative less-than-fraction by number of particles: F; cumulative less-than-fraction by mass (or volume) of particles: G. Alternatively the last two may easily be stated as cumulative-greater-than just as well. If a distribution formula is given in any one of these four forms, in principle the other three can be derived from it by the use of these relationships.

$$p = \frac{dF}{dd_p} \qquad q = \frac{dG}{dd_p} \tag{4) and (9}$$

$$F = \int_0^{d_p} p\,dd_p \qquad G = \int_0^{d_p} q \cdot dd_p \tag{18}$$

$$p = \frac{q}{d_p^3 \int_0^\infty \frac{q}{d_p^3}\,dd_p} \qquad q = \frac{d_p^3 \cdot p}{\int_0^\infty d_p^3 p\,dd_p} \tag{19}$$

$$F = \frac{1}{\int_0^\infty \frac{q}{d_p^3}\,dd_p} \int_0^d \frac{q}{d_p^3}\,dd_p \qquad G = \frac{1}{\int_0^\infty d_p^3 \cdot p\,dd_p} \int_0^d d_p^3 p \cdot dd_p \tag{20}$$

However, in some cases these conversions are not easy to perform in practice.

In order for a function to represent real data satisfactorily, it must usually possess the following properties:

$$\text{As } d_p \to 0,\ p = F = p = G \to 0$$

$$\frac{dp}{dd_p} = \frac{dF}{dd_p} = \frac{dq}{dd_p} = \frac{dG}{dd_p} = 0$$

$$\text{As } d_p \to \infty,\ p = q \to 0;\ F = G \to 1$$

$$\frac{dp}{dd_p} = \frac{dF}{dd_p} = \frac{dq}{dd_p} = \frac{dG}{dd_p} \to 0$$

Although, of course, there are no particles either of zero or of infinite size, there is usually a negligible error introduced by using a function which includes these limits, and such a function is usually of the simpler two-constant variety. In case it is important to represent either a minimum particle size significantly greater than zero, and/or to represent a known upper limit to the particle size, one of the more elaborate functions, involving three constants, is required. The limiting conditions are modified accordingly. In addition, p and q must each possess a maximum point corresponding to a point of inflection in F and G respectively. Functions possessing these characteristics will only be capable of describing uni-modal distributions.

These general statements, with a few exceptions as noted, may be applied to each of the functions given below. Each of these has been discussed in detail in standard references which are cited. The following treatment is devoted to a practical working summary of the characteristics of each function and the method of testing its fit to actual data.

1. Log Normal Probability [1, 2, 4,11]

This is considered to be the most generally applicable function to describe the aerosols encountered in the ambient air, and to describe many process ducts encountered in emission control problems. It is also relatively convenient to use, and for this reason may be used as a satisfactory approximation in some cases.

$$p(d_p) = \frac{1}{d_p \sqrt{2\pi} \ln \sigma_g} \exp - \left[\frac{\ln d_p/\bar{d}_p}{\sqrt{2}\,(\ln\sigma_g)}\right]^2 \qquad (21)$$

$$F(d_p) = \frac{1}{\sqrt{2\pi}\ \ln\sigma_g} \int_{-\infty}^{\frac{\ln d_p/\bar{d}_p}{\sqrt{2}\ \ln\sigma_g}} \exp - \frac{(\ln d_p/\bar{d}_p)^2}{2(\ln\sigma_g)^2}\ d\ \ln d_p \tag{22}$$

These expressions result from applying the Gaussian or normal law of distribution to $\ln d_p$ instead of d_p itself. Irani and Callis [4], and Mugele and Evans [10] have given derivations to indicate how this would be consistent with certain physical facts about aerosol and particle properties. The two constants are: $\bar{d}_p$ = number median particle size (NMD), for which F = 0.50 and σ_g = geometric standard deviation (dispersion indicator) $\geq$ 1. When σ_g = 1, the aerosol is said to be monodisperse, all particles the same size. The maximum value of p, corresponding to the inflection in F, occurs at

$$d_{p_{max}} = \bar{d}_p/\sigma_g^{\ \ln \sigma_g} \tag{23}$$

for which

$$p_{max} = \frac{\sigma_g \ln\sigma_g}{\bar{d}_p \sqrt{2\pi}\ \ln \sigma_g} \exp\left(-\frac{\ln^2 \sigma_g}{2}\right) \tag{24}$$

For ease in evaluation, F may readily be transformed into terms of the "error function" erf. Thus

$$f_{a-b} = F(d_{p_a}) - F(d_{p_b})$$

$$= \frac{1}{2}\left[\operatorname{erf}\left(\frac{\ln d_{p_a}/\bar{d}_p}{\sqrt{2}\ \ln\sigma_g}\right) - \operatorname{erf}\left(\frac{\ln d_{p_b}/\bar{d}_p}{\sqrt{2}\ \ln\sigma_g}\right)\right] \tag{25}$$

or

$$F(d_p) = \frac{1}{2} \left[1 + \text{erf} \left(\frac{\ln d_p/\bar{d}_p}{\sqrt{2}\,\ln\sigma_g} \right) \right] \tag{26}$$

Tables of erf are available in standard handbooks. Smith [12] gives a very good approximation formula which is well-suited for use with a pocket calculator. Flynn [13] shows how to substitute a series of straight lines to represent the error function in computer calculation.

It may be shown that the mass (or volume) distribution functions q and G are of the same forms as those for p and F and have the same value of σ_g. The only difference is that $\bar{d}_p$ (NMD) is replaced by $\bar{d}_{pm}$, the mass median size (MMD), for which G = 0.50. The interrelationship is:

$$\ln\bar{d}_{pm} = \ln\bar{d}_p + 3\ \ln^2\sigma_g \tag{27}$$

or

$$\ln \text{MMD} = \ln \text{NMD} + 3\ \ln^2\sigma_g$$

It is also possible to consider the distribution function in terms of the diameter, and of the surface, of the particles in addition to simply their number count or their volume (mass). In each case, this distribution law applies with the same value of σ_g, but a different value of the median size:

$$\ln\text{DMD} = \ln\text{NMD} + \ln^2\sigma_g \text{ (diameter median)} \tag{28}$$

$$\ln\text{SMD} = \ln\text{NMD} + 2\ \ln^2\sigma_g \text{ (surface median)} \tag{29}$$

These median values are always in the order MMD > SMD > DMD > NMD.

A general relationship exists among any of these medians and among the several mean values as defined in section II.D:

$$\ln\bar{D}_{m,w} = \ln\bar{d}_{p_x} + \frac{m + 2w - 2x}{2} \ln^2\sigma_g \tag{30}$$

Here $\bar{D}_{m,w}$ = mean value, as defined by m and w in Equation (10)

$\bar{d}_{p_x}$ = median value, based upon

x = 0 for number cound $\bar{d}_{p_o}$ = NMD

x = 1 for diameter $\bar{d}_{p_1}$ = DMD

x = 2 for surface $\bar{d}_{p_2}$ = SMD

x = 3 for volume (mass) $\bar{d}_{p_3}$ = MMD

The easiest way to test data for conformity with this function an and, at the same time to obtain values for σ_g and the appropriate median, is to use a special graph paper commercially available, which is known as log-probability paper. One axis is a standard log scale, usually of 2 cycles, used for plotting log d_p. The other axis is a specially constructed scale corresponding to the error function integral as in (26), and used for plotting cumulative-percent-less-than on whatever basis is desired, F or G. An illustration is shown in Figure 7.

Whenever experimental data points are plotted on this paper, a straight line will be obtained if the data agree with this distribution function. The slope of this line is determined by the value of σ_g. This is easily calculated by reading the values of d_p corresponding to $+\ \sigma_g\bar{d}_p$ and $-\ \sigma_g\bar{d}_p$. From (24) it may be shown that $+\ \sigma_g\bar{d}_p$ is located at 84.13%-less-than, and $-\ \sigma_g\bar{d}_p$ at 15.87%-less-than. These values, designated as $d_{p_{84}}$ and $d_{p_{16}}$ respectively, are used together with the median size $d_{p_{50}}$ to calculate σ_g:

$$\sigma_g = \sigma\bar{d}_p/\bar{d}_p = d_{p_{84}}/d_{p_{50}} = d_{p_{50}}/d_{p_{16}} = \sqrt{d_{p_{84}}/d_{p_{16}}} \tag{31}$$

The straight lines on the log-probability plot corresponding to each of the four kinds of distribution (count, diameter, surface, mass) will be parallel to each other and each will have the same value of σ_g. The spacing between them will be determined by the median values $\bar{d}_{p_x}$, as given above in (27), (28) and (29).

a. Example 4. The table gives data on the cumulative size distribution of a sample of glass beads as quoted by Irani and Callis [4]. Test whether these data follow the log-normal probability function and if so, determine σ_g, NMD, MMD and the Sauter mean diameter ($\bar{D}_{1,2}$). Show how to use the constants found to calculate the frequency distribution in 5-micron intervals.

Size Distribution of a Sample of Glass Beads [4]

	Cumulative Distribution			
	Weight Distribution		Number Distribution	
Size (microns)	% Cumulative (Greater than)	G % Finer than	% Cumulative (Greater than)	F % Finer than
5	100.0	0.0	100.00	0.0
10	99.9	0.1	99.0	1.0
15	98.4	1.6	86.2	13.8
20	89.5	10.5	58.0	42.0
25	71.5	28.5	32.0	68.0
30	50.0	50.0	15.0	85.0
35	32.1	67.9	7.0	93.0
40	19.2	80.8	2.8	97.2
45	10.8	89.2	1.2	98.8
50	6.0	94.0	0.5	99.5
55	3.0	97.0	0.15	99.85
60	1.9	98.1	0.09	99.91
100	0.0	100.0	0.0	100.0

If the raw data points are plotted on regular graph paper, the curves will be similar to those of Fig. 6. However, it is better to plot them on the special log-probability paper as shown in Fig. 7. A straight line is clearly defined and drawn through each set of points. The following values are read from these lines;

	F-Number Distribution	G-Weight Distribution
d_{p84} - μm	29.7	41.7
d_{p50} - μm	21.5 (NMD)	30 (MMD)
d_{p16} - μm	15.5	21.7

Number Distribution (31), F line: $\sigma_g = 29.7/21.5 = 1.38$ or $\sigma_g = 21.5/15.5 = 1.39$, or $\sigma_g = \sqrt{29.7/25.5} = 1.38$. Since the values of σ_g are the same for each line, the lines are shown to be parallel, as indeed they appear to be to the eye.

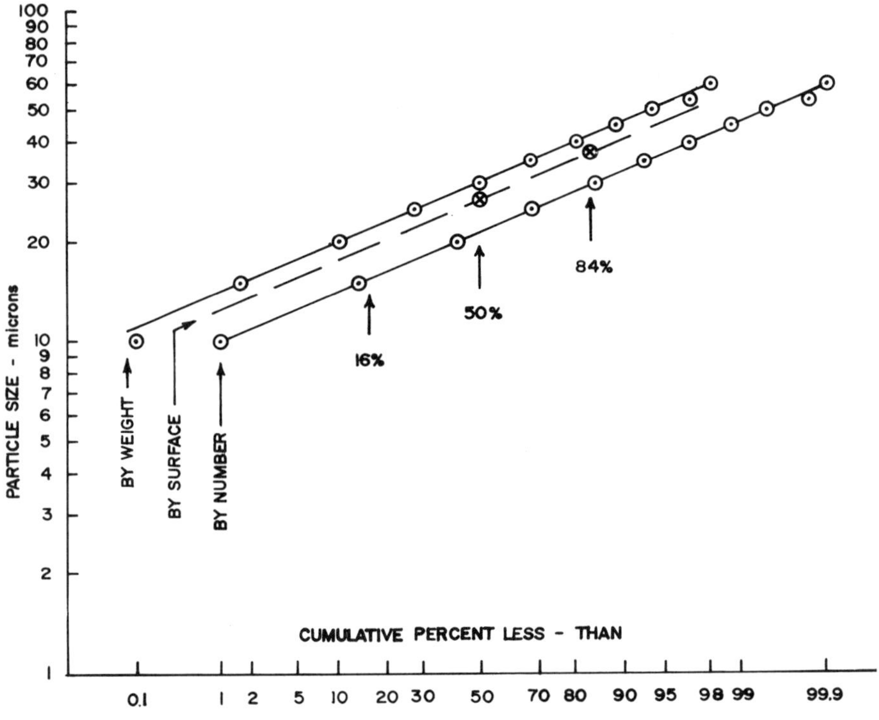

FIG. 7. Log-probability size distribution, Example 3: sample of glass beads [4] ; MMD = 30 microns; NMD = 21.5 microns; $\sigma_g = 1.38$.

Relationship between medians (27): solving for

$$\ln\sigma_g = \frac{\ln \text{MMD/NMD}}{3} = \frac{\ln 30/21.5}{3} = 0.3332; \quad \sigma_g = \underline{\underline{1.395}}.$$

This value

checks very well with those obtained from the slope of each line. Taking $\sigma_g = 1.39$ the Sauter mean diameter is calculated from (30), using m = 1, w = 2, x = 3, as $\ln \bar{D}_{1,2} = \ln 30 + \frac{1 + 2(2)-2(3)}{2} \ln^2 1.39$;

Sauter mean: $\bar{D}_{1,2} = \underline{\underline{28.4}}$ μm.

If desired, lines could also be constructed on Fig. 7 to give the distribution by diameter, or by surface. They would be drawn parallel to the count and mass lines, i.e. with $\sigma_g = 1.39$, and pass through a median size as calculated from (28) or (29). For example, the surface median size would be: $\ln \text{SMD} = \ln 21.5 + \ln^2 1.39$; SMD = 26.7 μm. $d_{p_{84}} = 26.7 \times 1.39 = 37.1$ μm. This line is shown dashed on Fig. 7.

To construct the number frequency distribution in 5-μm intervals, (3) or (23) is used. The integrals are approximated by using the value of p at the center of the interval:

$$f_{0-5} = \int_0^5 p \cdot dd_p \approx p_{2.5}(5-0) \text{ or } g_{0-5} = \int_0^5 q \cdot dd_p \approx q_{2.5}(5-0).$$

The values of p, or q, are calculated from (21). To cover the range of sizes from 0 to 60 μm, twelve intervals of 5 μm each are used, and a final interval is taken from 60 to 100 μm because this represents a small portion of the total. The results are tabulated for both number size-distribution f_i and mass size-distribution g_i. An alternate approach is to calculate f from the erf relationship given by (23). This involves no approximation. The results of both methods are tabulated for the number and mass frequency. They are seen to agree well with each other and with the original data cited by Irani and Callis, except for the upper end of the range where the original data show an unexpected increase in occurrence of particles. Figure 8 is a plot of these results.

The method of calculation for each column in the table is shown below and keyed to the table by the letters (A), (B), etc., appearing at the head of the column.

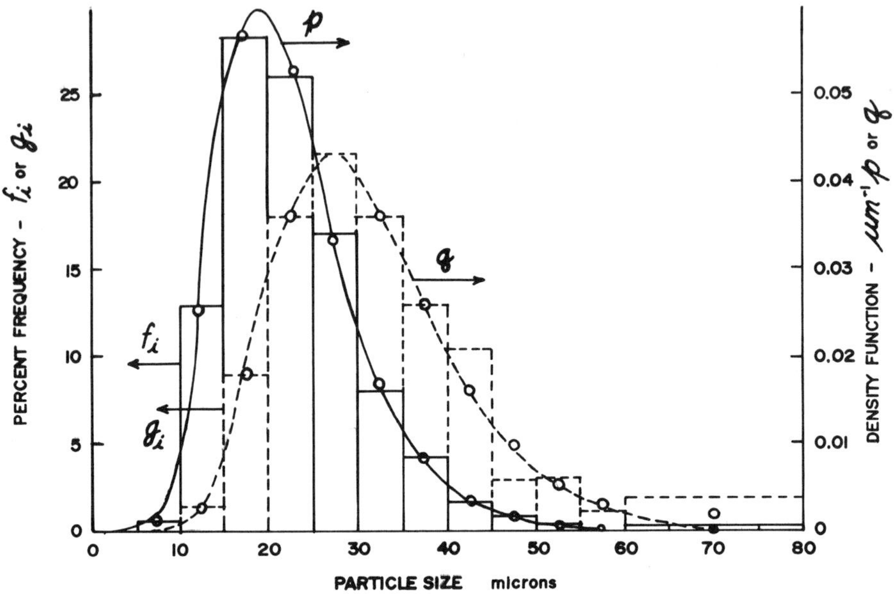

FIG. 8. Density and frequency distribution, Example 3: sample of glass beads 4. Solid lines, by number, dashed lines, by mass.

$$\text{(A)}\quad p_{1/2} = \frac{1}{d_{p_{1/2}}\sqrt{2\pi}\,\ln 1.39}\ \exp - \left[\left(\frac{\ln d_{p_{1/2}}/21.5}{\sqrt{2}\,\ln 1.39}\right)\right]^2 \qquad (21)$$

$$\text{(B)}\quad f_i \simeq p_{1/2}\,(d_{p_a} - d_{p_b}) \qquad (3)$$

$$\text{(C)}\quad f_i = F_a - F_b = \frac{1}{2}\left[\operatorname{erf}\left(\frac{\ln d_{p_a}/21.5}{\sqrt{2}\,\ln 1.39}\right) - \operatorname{erf}\left(\frac{\ln d_{p_b}/21.5}{\sqrt{2}\,\ln 1.39}\right)\right] \qquad (23)$$

$$\text{(D)}\quad q_{1/2} = \frac{1}{d_{p_{1/2}}\sqrt{2\pi}\,\ln 1.39}\ \exp - \left[\left(\frac{\ln d_{p_{1/2}}/30}{\sqrt{2}\,\ln 1.39}\right)\right]^2 \qquad (21)$$

(E) $g_i \simeq q_{1/q} (d_{p_a} - d_{p_b})$ (3)

(F) $$g_i = G_a - G_b = \frac{1}{2}\left[\text{erf}\left(\frac{\ln d_{p_a}/30.0}{\sqrt{2}\ \ln 1.39}\right) - \text{erf}\left(\frac{\ln d_{p_b}/30.0}{\sqrt{2}\ \ln 1.39}\right)\right] \qquad (23)$$

Calculation for maximum points:

$$p_{max} = \frac{1.39^{\ln 1.39}}{21.5\sqrt{2\pi}\ \ln 1.39} \exp\left(-\frac{\ln^2 1.39}{2}\right) = 0.0595 \qquad (24)$$

$$d_{p_m} = 21.5/1.39^{\ln 1.39} = 19.3\ \mu m \qquad (23)$$

$$q_{max} = \frac{1.39^{\ln 1.39}}{30\sqrt{2\pi}\ \ln 1.39} \exp\left(-\frac{\ln^2 1.39}{2}\right) = 0.0426 \qquad (24)$$

$$d_{p_m} = 30/1.39^{\ln 1.39} = 26.9\ \mu m \qquad (23)$$

2. Modified "Upper Limit" Log-Normal [10]

This function was devised by Mugele and Evans in order to represent the size distribution in sprays and other aerosols having a similar mechanism of formation. It incorporates a parameter to represent the maximum stable drop-size $d_{p_{max}}$ as the upper-limit to the particle distribution. It assumes then that the ratio $u = d_p/(d_{p_{max}} - d_p)$ should be distributed log-normally, as d_p is in (1) above. This results in a three-constant equation: $d_{p_{max}}$, σ_g, and u_{50}(median).

To test data for a fit to this function the special log-probability graph paper is used to determine whether a value of $d_{p_{max}}$ can be found such that a plot of log u vs. weight (or volume) - cumulative-per cent-less-than, G, on probability scale will yield a straight line. The value of $d_{p_{max}}$ may be determined by trial-and-error, or it may be estimated, at least for a first trial, from:

Results of Example 4

Size Interval $(d_{pb}-d_{pa})$ -μm	Midpoint of Interval $-d_{p1/2}$	(A) $p_{1/2}$	(B) f_i	(C) f_i	Exptl f_i	(D) $q_{1/2}$	(E) g_i	(F) g_i	Exptl g_i
0-5	2.5	$2x10^{-10}$	10^{-9}	0	0	$2x10^{-13}$	$1x10^{-12}$	0	0
5-10	7.5	$1x10^{-3}$	0.005	0.010	0.010	$2x10^{-5}$	$1x10^{-4}$	0	0.001
10-15	12.5	0.0249	0.125	0.127	0.128	0.0028	0.014	0.017	0.015
15-20	17.5	0.0569	0.285	0.279	0.282	0.0181	0.091	0.092	0.089
	(19.3) Max	(0.0595)							
20-25	22.5	0.0533	0.267	0.260	0.260	0.0368	0.184	0.181	0.180
	(26.9) Max					(0.0426)			
25-30	27.5	0.0333	0.167	0.168	0.170	0.0425	0.213	0.210	0.215
30-35	32.5	0.0170	0.085	0.086	0.080	0.0362	0.181	0.180	0.179
35-40	37.5	0.0078	0.039	0.040	0.042	0.0257	0.129	0.129	0.129
40-45	42.5	0.0033	0.017	0.017	0.016	0.0163	0.082	0.082	0.104
45-50	47.5	0.0014	0.007	0.007	0.007	0.0096	0.048	0.049	0.028
50-55	52.5	0.0006	0.003	0.003	0.0035	0.0054	0.027	0.028	0.030
55-60	57.5	0.0002	0.001	0.001	0.0006	0.0030	0.015	0.015	0.011
60-100	80	$5x10^{-6}$	$2.5x10^{-5}$	0.0009	0.0009	0.00018	0.0072	0.016	0.019
	Totals		1.001	0.999	1.000		0.991	0.999	1.000

COMMENTS ON RESULTS OF EXAMPLE 4

1. The numbers listed in columns headed (A), (B), (C), (C), (E), (F) were calculated by the corresponding equations listed on pages 48 and 49.

2. The values in column (A) are plotted on Figure 8, solid curve, those of column (D) on Fig. 8, dashed curve.

3. Note that the smooth curves for p and q are not based upon the histograms for f_i and g_i, but rather upon the calculations for F and G.

4. The calculated value of f_i in columns (B) and (C) are to be compared with each other and with the experimental values. The agreement is generally quite good.

5. The calculated values of g_i in columns (E) and (F) are to be compared with each other and with the experimental value. This agreement is also quite good.

6. The raw data (page 45) show an increase in the number and the mass of particles which would be expected in the upper end of the range (above 60 microns) according to the log-probability distribution.

$$\frac{d_{p_{max}}}{d_{p_{50}}} = \frac{d_{p_{50}}(d_{p_{90}} + d_{p_{10}}) - 2\, d_{p_{90}}\, d_{p_{10}}}{d_{p_{50}}^{2} - d_{p_{90}}\, d_{p_{10}}} \tag{32}$$

Here $d_{p_{10}}$, $d_{p_{50}}$, and $d_{p_{90}}$ represent values of particle size at 10%, 50% (median), and 90%-finer-than as read from an initial plot of d_p vs. G.

Having found a straight line on this plot, the mass median value of u, $u_{50} = d_{p_{50}}/(d_{p_{max}} - d_{p_{50}})$ at G = 50%, is read. Then from the values of u_{50} and u_{84} (also read from the straight line), the value of σ_g is found:

$$\sigma_g = u_{84}/u_{50} \tag{33}$$

Finally, the distribution functions are:

$$q = \frac{u_{50}}{\sqrt{2\pi}\, \ln \sigma_g} \exp - \left[\left(\frac{\ln u/u_{50}}{\sqrt{2}\, \ln \sigma_g}\right)\right]^2 \tag{34}$$

and

$$G = \frac{1}{\sqrt{2\pi}\, \ln \sigma_g} \int_{-\infty}^{\ln u/u_{50}} \exp - \left[\left(\frac{\ln u/u_{50}}{\sqrt{2}\, \ln \sigma_g}\right)\right]^2 d(\ln u) \text{ or}$$

$$G = \frac{1}{2}\left[1 + \operatorname{erf} \frac{\ln u/u_{50}}{\sqrt{2}\, \ln \sigma_g}\right] \tag{35}$$

It would seem that there is no reason why similar functions could not be found in terms of number distribution (p and F) as well, noting, however, that the upper limit of the integrals in (18), (19) and (20) must be $d_{p_{max}}$.

Mugele and Evans tested this "upper-limit distribution" against a variety of data on sprays and found it to give a better fit in each case than the log-probability, the Rosin-Rammler, and the Nukiyama-Tanasawa distributions. The last two are described below. An additional test is given in Example 5.

a. Example. Kim and Marshall [14], in a study of the production of sprays from a pneumatic atomizer, determined a generalized drop-size distribution by weight as given in the first two columns of the table below. Here, the drop-size d_p^* is a normalized value obtained by dividing the actual size in each case by the mass-median diameter. Test these data against the upper-limit function, and establish the maximum drop size.

Since d^*_{P10}, d^*_{P50} and d^*_{P90} are all actual entries in the table, the value of $d_{P_{max}}$ may be estimated at once: $d^*_{P10} = 0.33$; $d^*_{P50} = 1.0$; $d^*_{P90} = 2.0$.

$$\frac{d^*_{P_{max}}}{1.0} = \frac{1.0(2.0 + 0.33) - 2(2.0 \times 0.33)}{(1.0)^2 - (2.0 \times 0.33)} = 29.7 \tag{32}$$

Using this value, the entries for u in the third column of the table are calculated. These points are then plotted against G on the log-probability graph Figure 9. A good straight line is obtained, confirming the value of $d^*_{P_{max}}$ and a fit to the upper-limit equation.

From the straight line $u_{50} = 0.52$, $u_{84} = 1.53$ and $\sigma_g = 1.53/0.52 = 2.94$. Then q (if desired) and G may be calculated from

$$q = \frac{0.52}{u\sqrt{2}\ln 2.94} \exp - \left[\left(\frac{\ln u/0.52}{\sqrt{2}\ln 2.94} \right) \right]^2 \tag{34}$$

$$G = \frac{1}{2}\left[1 + \operatorname{erf} \frac{\ln u/0.52}{\sqrt{2}\ln 2.94} \right] \tag{35}$$

The calculated values of G are listed in fourth column and are seen to reproduce the original values very well.

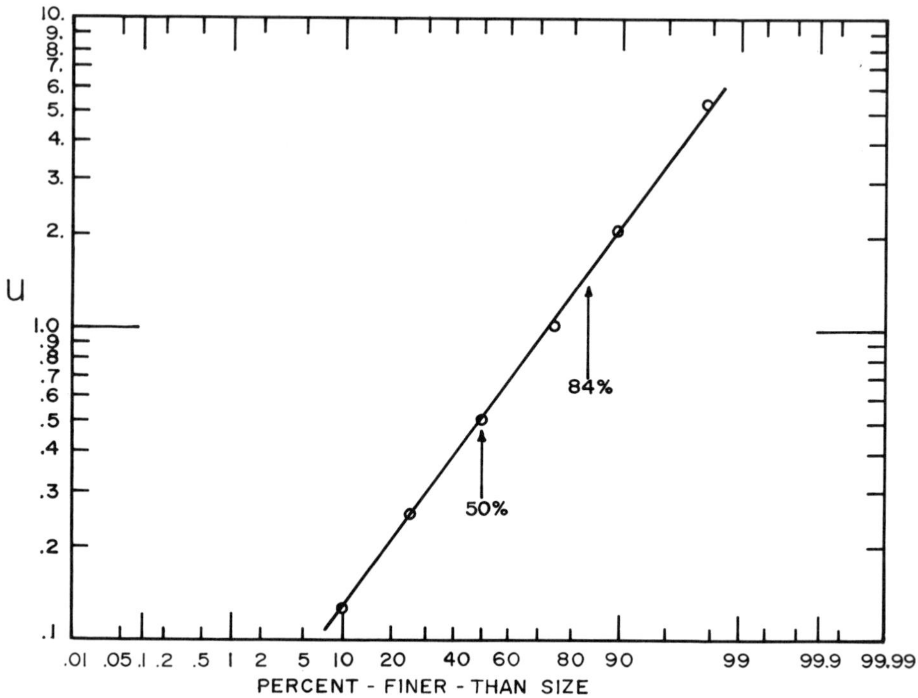

FIG. 9. Test of upper-limit function. Example 5: data on sprays [14]; $u = d_p^*/d_{p_{max}}^* - d_p^*)$, $d_{p_{max}}^* = 29.7$ microns $u_{50} = 0.508$ $\sigma g = 2.94$. Reproduced by permission from American Institute of Chemical Engineers Journal.

G Exp't'l	d_p^*	$u = d_p^*/(2.97 - d_p^*)$	G Calculated
0.10	0.33	0.125	0.093
0.25	0.60	0.253	0.252
0.50	1.0	0.508	0.491
0.75	1.5	1.02	0.734
0.90	2.0	2.06	0.899
0.98	2.5	5.32	0.984

3. Weibull Distribution [15, 16, 17]

A distinctly different and wholly empirical equation, so far as particle size distribution is concerned, is the Weibull function. Originally derived and widely used for other purposes [15], its applicability has been demonstrated by Steiger [16] for describing size distribution in various kinds of dusts.

It is a three-constant equation stated basically in the cumulative form:

$$F(d_p) = 1 - \exp = \frac{(d_p - \gamma)^\beta}{\alpha} \tag{36}$$

The utility of this form lies in the constant γ, which may be taken to represent the size of the smallest particle present, in those cases when it is imperative or simply more accurate to describe a distribution without letting $d_p \to 0$. Furthermore, α may be expressed as a "special" particle size $\bar{d}_p$, raised to the β power. Rewriting (36),

$$F(d_p) = 1 - \exp \left[\left(\frac{d_p - d_{p_{min}}}{\bar{d}_p} \right) \right]^\beta = 1 - \exp - X^\beta \tag{37}$$

it is seen how the ratio $X = (d_p - d_{p_{min}})/\bar{d}_p$ plays a role somewhat analogous to that of u in the "upper-limit" form, as in (2) above. Notice, however, that $\bar{d}_p$ may not be a particle size actually present, for whenever $\alpha^{1/\beta} < \gamma$, $\bar{d}_p$ will be smaller than $d_{p_{min}}$. The third constant β is a measure of the degree of dispersion of the sizes present.

The equation may be discussed and its properties demonstrated either in the form of (36) or the generalized form of (37). The population density is derived using (17), as

$$p(d_p) = \frac{\beta}{\alpha} (d_p - \gamma)^{\beta - 1} \exp - \frac{(d_p - \gamma)^\beta}{\alpha} \tag{38}$$

or

$$p(X) = \frac{\beta X^{\beta-1}}{\bar{d}_p} \exp - X^{\beta} \tag{39}$$

Examples of the graphs which the Weibull equations may describe are best shown in generalized form, as in Figure 10 (cumulative) and Figure 11 (density). It is seen that provided $\beta > 1$ a variety of S-shaped curves may be represented by the form of (36), and that the forms given by (38) are not unlike the skewed type of probability curves represented by the log-normal function (21). Steiger [16] states that when $\beta = 3.25$, most of the Weibull curve is identical with the normal (not log-normal) distribution function. Usually β will lie between 1.0 and 3.0.

The values of α, β, and γ may be found from a set of data by a graphical technique utilizing a linearized plot. For this purpose, plot log ln $1/(1-F)$ vs. log $(d_p - \gamma)$, for from (36)

$$\ln 1/(1 - F) = \beta \log (d_p - \gamma) - \log\alpha \tag{44}$$

That is, plot ln $1/(1-F)$ vs. $(d_p - \gamma)$ on log-log coordinate scales. There is also a special graph paper available for this purpose [18]. If the value of γ cannot be seen at once from the data, i.e. $\gamma = d_{p_{min}}$ for which $f = F = 0$, then several values of γ will have to be tried until one is found that does give a straight line on such a plot. If none can be found, the Weibull function does not fit the data.

4. Rosin Rammler Distribution [19]

Of lesser importance for aerosol work is this empirical relationship, originally proposed to represent size distribution of relatively coarse dusts produced by grinding of solids. It is basically of the two-constant form.

$$G = 1 - \exp - ad_p^s \tag{45}$$

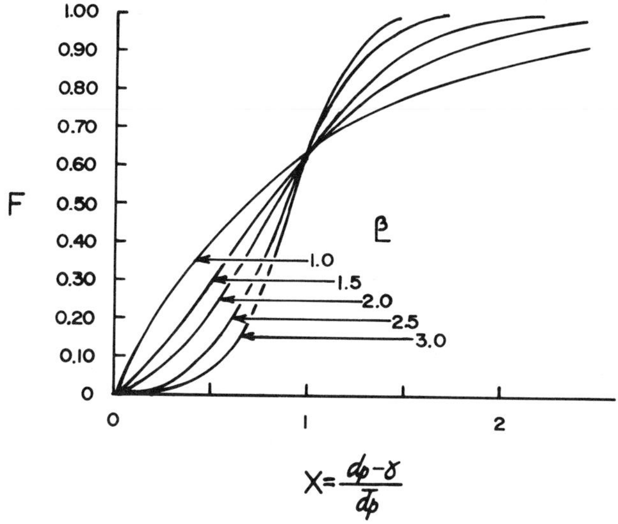

FIG. 10. Examples of Weibull cumulative plots in generalized form. See equation (39).

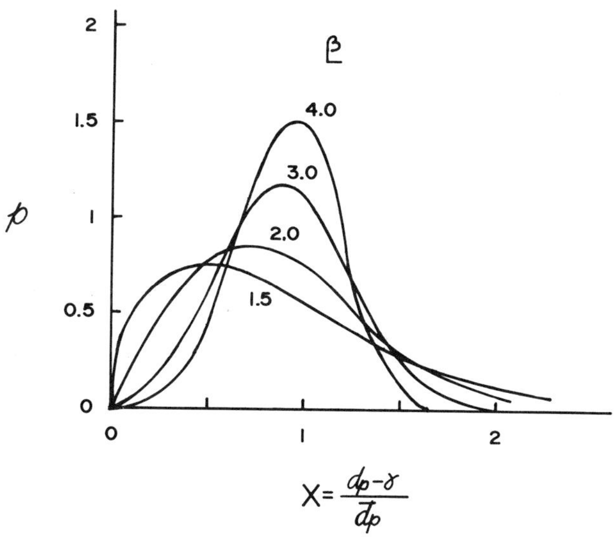

FIG. 11. Examples of Weibull density plots in generalized form. See equation (39).

which is seen to be a special case of the Weibull (36) in which $\alpha = 1/a$, $\beta = s$ and $\gamma = 0$. A mean size $\bar{d}_p = 1/a^{1/s}$, and the function written in the alternate form

$$G = 1 - \exp - (d_p/\bar{d}_p)^s \tag{46}$$

Data may be tested for a fit to this form in a manner similar to that used with the Weibull. Thus (44) becomes:

$$\log \ln 1/(1-G) = s \log d_p + \log a \tag{47}$$

The same kind of special paper may be used.

5. Roller Distribution [20]

Another two-constant empirical formula, said to be applicable over a wide range of particle sizes of powdered industrial materials (but probably of limited value in describing aerosols) is given as:

$$G = a d_p^{1/2} \exp(- s/d_p) \tag{48}$$

A severe limitation of the use of this function is that it does not fulfill one of the basic conditions, namely that $d_p \to \infty$, $G \to 1$. Instead, as $d_p \to \infty$, $G \to \infty$, and $G\sqrt{d_p} \to a$. Where $G = 1$, d_p has a value d_{p100} in accord with

$$\frac{1}{a\sqrt{d_{p100}}} = \exp - \left(\frac{s}{d_{p100}}\right) \tag{49}$$

The weight-density function is

$$q = a\sqrt{d_p} \exp\left[\left(-\frac{s}{d_p}\right)\right]\left[\left(\frac{1}{2d_p} + \frac{s}{d_p^2}\right)\right] = G\left[\left(\frac{1}{2d_p} + \frac{s}{d_p^2}\right)\right] \tag{50}$$

which likewise does not approach zero as $G \to 1$.

Great care must, therefore, be taken in using this function, not to apply it in the region of sizes near that for which $G = 1$. For this reason, it does not seem to warrant more detailed study.

Data may be tested for accord with it by plotting $\log (G/\sqrt{d_p})$ vs. $1/d_p$, for

$$\log G/\sqrt{d_p} = \log a - 0.4343\ s/d_p \tag{51}$$

If a straight line is obtained its slope will be equal to $-0.4343\ s$, and its intercept is log a.

6. Nukiyama-Tanasawa Distribution [21]

This expression was offered in the population-density (p) form to express the size distribution in mechanically dispersed mists, formed by atomization:

$$p = a\ d_p^{\ 2} \exp\ (-\ bd_p)^s \tag{52}$$

This appears to have three constants a, b, s, but they are not independent. For, in order to fulfill the requirement of (4), an interrelationship

$$\frac{a\ b^{-3/s}\Gamma(3/s)}{s} = 1 \tag{53}$$

must be fulfilled, where Γ indicates the gamma function or generalized factorial. Therefore, only two of the constants are independent.

Through application of (19), the corresponding weight-density function is found to be

$$q = \frac{sb^{6/s}}{\Gamma(b/s)}\ d_p^{\ s} \exp - bd_p^{\ s} \tag{54}$$

The cumulative forms for F and G may be found by application of (20).

If data are available for p, a test of fit to (52) may be made by plotting $\ln p/d_p^2$ vs. d_p^s to see if a straight line may be determined. For from (52) and (53) combined:

$$\ln p/d_p^2 = \ln sb^{3/s}/\Gamma(3/s) - b\, d_p^s \tag{55}$$

Values of s must be tried e.g. as 1/2, 1/3, 1/4—to see whether one can be found to produce a straight line. If so, then -b is the slope of that line. Then the values of b and s so determined may be checked to see whether $\ln sb^{3/s}/\Gamma(3/s)$ agrees with the measured intercept of the line. An analogous procedure could be carried out with data for q using the linearized form of (54).

The value of this function has been shown by Mugele and Evans [10] to be somewhat limited. They found that the upper-limit form of the log-normal function fitted data for atomized sprays better. Furthermore, the information on pneumatic atomizers has been greatly extended and improved by Kim and Marshall [14] whose generalized drop size distribution function has been cited in Example 5. This may be considered to supercede the Nukiyama-Tanasawa function.

7. Multi-Modal Distributions

If the plot of F (or G) vs. d_p on log-probability paper is not straight, the reason may be that the distribution has more than one mode. Each mode is identified by a peak value of p (or q) and corresponds to a point of inflection on the F (or G) plot. It may be that there are two or more "parent" distributions, each of which is log-normally distributed, mixed together in such proportions as to produce a multi-modal composite. Examples of this are given by Allen [1], one of which is shown in Figure 12.

In order to magnify the presence of multimodes it is useful to

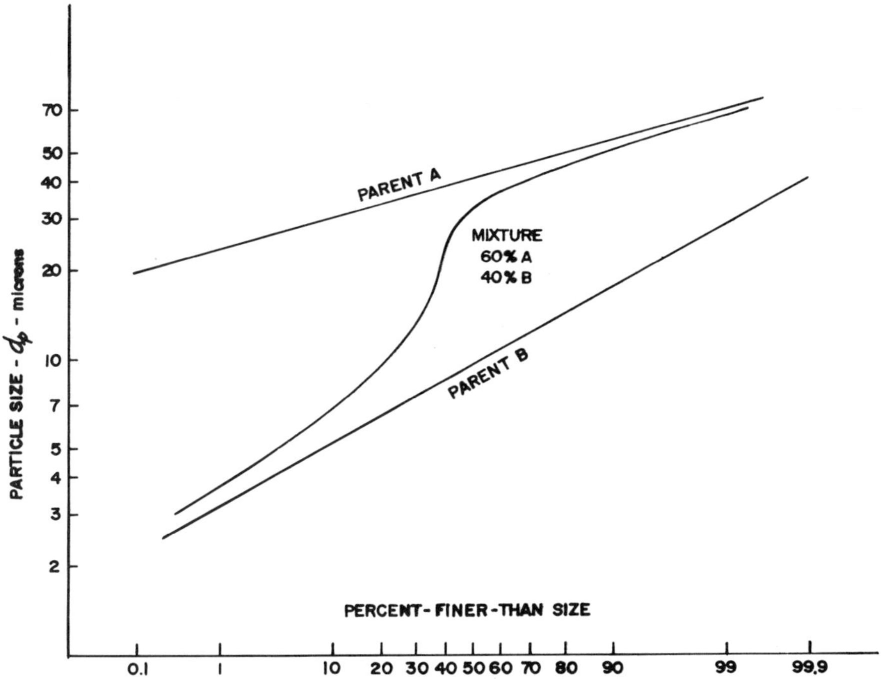

FIG. 12. A bimodal distribution which is a mixture of log-probability parent distributions.

plot data in terms of $dn/d \log d_p$ vs. $\log d_p$ or for discrete intervals $\Delta n/\Delta \log d_p$ vs. $\log d_p$. This is equivalent to $d_p \times dn/dd_p$ or $p \cdot d_p$ vs. $\log d_p$.

One of the most important examples of a multi-modal distribution is that of the particles in the usual ambient urban atmosphere. In terms of mass measurements (g) two peaks and often three are observed. These have been distinguished according to characteristic sources of origins of particles in each mode. The table below from Willeke and Whitby [22] summarizes these facts.

Names	Transient nuclei range	Accumulation range	Mechanical aerosol range
Size 0.001 -microns |	0.01 1 | 2	0.1 1 | 3 |	10 100 4 | 5 |
Sources	Combustion Heterogeneous nucleation	Coagulation from transient nuclei Condensation Combustion	Windblown dust Large particle emissions Sea spray
Lifetime	< 1 hour	Days	Hours Minutes Days Hours

C. Note of Caution

In performing the calculations explained above, one must not lose sight of the accuracy and precision of the raw data upon which they are based. Particle size measurements, counting and weighing, are difficult techniques at best. It requires painstaking care, and close attention to the principles of the method of measurement, in order to obtain reliable, meaningful and reproducible values.

Elaborate calculations of density functions and mean particle sizes, are not worthwhile unless the raw data are reliable and well-determined. Consequently, the manipulator of such data must have a good understanding of the problems and techniques of the measurer who obtained the data. In deciding what calculations to undertake, one must bear in mind the purpose for which the data are to be used. In fact this should be the primary consideration when the method of measurement is selected at the outset.

For example, when a cumulative distribution by mass is considered, little attention need be given to the particle sizes corresponding to either the upper two-or three-percent or the lower two-or three-percent of the aerosol as those will represent a relative small proportion of the total mass. A size-distribution function which fits quite well through the mid-range of sizes may be quite acceptable in describing the aerosol, say for purposes of calculation of a dust collector performance on the mass basis. On the other hand, the principle concern may be with the number of particles in the finest size-range because of the health hazard they represent. This might call for emphasis upon a different kind of distribution function in order to represent this range most accurately.

REFERENCES

1. T. Allen, *Particle Size Measurement*, Wiley (Halstead Press), London, 2nd Edition, 1975.

2. C. Orr and J. M. Dallavalle, *Fine Particle Measurement*, Macmillan, New York, 1959.

3. L. D. Carver, *Indus. Res.:* 44 (Aug., 1971).

4. R. R. Irani and C. F. Callis, *Particle Size: Measurement, Interpretation and Application*, Wiley, New York, 1963.

5. W. C. McCrone and J. G. Delly, *The Particle Atlas*, Ann Arbor Science, Ann Arcor, 3rd Edition, 1977.

6. D. Kunii and O. Levenspiel, *Fluidization Engineering*, Wiley, New York, 1969.

7. P. M. Giever in *Air Pollution* (A. C. Stern, Ed.) 3rd Edition, Vol. 3, Academic Press, New York, 1976.

8. N. R. Iammartino, *Chem. Eng.*, 32 (July 10, 1972).

9. K. F. Whitby and B. H. Y. Liu in *Encyclopedia of Chemical Technology* (H. F. Mark, J. J. McKetta, Jr. and D. F. Othmer, Ed.) 2nd Edition, Vol. 7, 430, Wiley (Interscience), New York, 1965.

10. R. A. Mugele and H. D. Evans, *Ind. Eng. Chem.*, 43: 1317 (1951).

11. L. Silverman, C. E. Billings and M. W. First, *Particle Size Analysis in Industrial Hygiene*, Academic Press, New York, 1971.

12. J. M. Smith, *Scientific Analysis on the Pocket Calculator*, Wiley (Interscience), New York, 1975.

13. B. P. Flynn, *Chem. Eng.*, 150 (Feb. 19, 1973).

14. K. Y. Kim and W. R. Marshall, *A.I.Ch.E.J.* 17: 581 (1971).

15. W. Weibull, *J. Appl. Mech.* 18: 293 (1951).

16. F. H. Steiger, *Chem. Tech.* 1: 225 (1971).

17. C. C. Harris, *Chem. Tech.* 1: 446 (1971).

18. C. C. Harris, *Powder Tech.* 5: 40 (1971/72).

19. P. Rosin and E. Rammler, *Koll. Zeit.* 67: 16 (1934).

20. N. Roller, *J. Franklin Inst,* 223: 609 (1937).

21. S. Nukiyama and Y. Tanasawa, *Trans. Soc. Mech. Engrs. (Japan),* 5: 63 (1939).

22. K. Willeke and K. T. Whitby, *J. Air Poll. Control Assoc.* 25: 529 (1975).

CHAPTER 3

GENERAL CONCEPTS OF PARTICULATE COLLECTION

I. MECHANISMS FOR COLLECTION

In order for suspended aerosol particles to be removed from a gas (i.e. collected) the gas must be passed through a zone in which the particles come under the influence of some kind of force (or forces) which causes them to be diverted from the flow direction of the stream. They must remain under the influence of the collecting force(s) a sufficient length of time to be diverted to contact some collecting surface where they are removed from the stream. In some cases, the initial path of the particle is such that it collides directly with the collector surface unaided by any force. This is referred to as direct interception.

The hopeful assumption is made that once a particle collides with the collecting surface, it adheres to it and is not subsequently reentrained into the gas stream. Provision must be made to maximize the validity of this assumption, as well as to remove the collected particles from the collecting surface in order that it may continue to function properly.

The basic situation may be illustrated in one of the sketches of Figure 1. The nature of the collecting force may be illustrated by the designated locations: gravitational, at A or C; centrifugal, at A; inertial, at B; direct interception, at D; diffusional, at C

(Brownman, eddy); electrostatic, at A or C; thermal, at A or C; and possibly others such as sonic, diffusiophoretic, photophoretic, etc. The agglomeration of small particles into large ones by the action of some of these forces (diffusional, electrostatic, sonic, etc.) may also aid the collection process. Several modes of collection may occur simultaneously.

There are two general categories of collecting surfaces as shown in Figure 1. One is the boundary surface of a containment zone through which the air is made to flow. This type of surface may be a plane wall or the inside of a hollow cylinder, as in A. The other is the outside of a target such as a cylinder (fiber) or sphere (drop of liquid) as in B and C, or some irregular shape such as that presented by the particles in a granular packing, or the fibers of a media filter. A target type of collector consists of an array or assembly of many target elements, as in a filter made up of a mat of fibers, or in a scrubber consisting of a stream of suspended droplets. Direct interception upon previously deposited particles may be the principal mode of collection by fabric filters.

The several types of collectors in common use may be classified according to the nature of the collecting force and collecting surface as shown in the following table:

Collector	Primary Collecting Forces	Collecting Surface
	Surface Collectors	
Settling chamber	Gravitational	Plane
Momentum	Gravitational Inertial	Plane or cylindrical
Cyclone	Centrifugal	Cylindrical
Electrostatic Precipitator	Electrostatic	Plane or cylindrical
Thermal collector[a]	Thermophoresis	Plane
Impinger[a]	Inertial	Plane
	Target Collectors	
Filter (media)	Inertial Diffusional Direct interception Electrostatic	Cylindrical fiber or granular

Filter (fabric)	Direct interception	Layer of particles. irregular
	Target Collectors	
Scrubber	Inertial Diffusional Direct interception	Spherical or irregular

[a]Usually used only for collection of small samples of aerosol.

In order to analyze or predict the performance of a collector, it is necessary to know how a given type of force acts upon a particle. The magnitude and direction of the force on a particle of given density and size must be predictable, so that a particle trajectory can be determined. This trajectory is a resultant of the collecting velocity

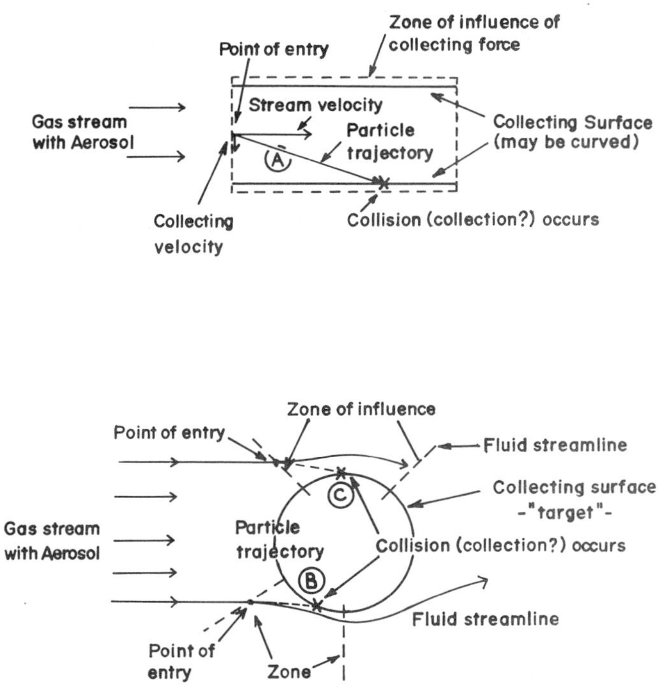

FIG. 1. Operation of collection forces within a zone of influence. (a) Boundary surface of a containment zone, (b) outside surface of a target zone.

set up by the force, and the gas stream velocity with which the particle enters the zone of influence. The motion of the particle in the direction of collection is always opposed by the frictional drag exerted by the gas on the particle. This must also be calculated and reckoned into the determination of the particle trajectory.

The interaction of these forces and velocities determines the differential equation which represents the net motion of the particle. The extent of our knowledge and ability to set up and solve such differential equations is treated in Chapters 4 and 5. The application of these results to the mathematical modelling of the collection performance of various kinds of collectors is the subject matter of Chapters 6, 7, 8, and 9.

II. EXPRESSING THE EFFECTIVENESS OF COLLECTION

The success of a particulate collection operation may be expressed either in terms of the amount of aerosol removed from the air stream, or in terms of the amount permitted to remain in it. Several different kinds of measures of collection effectiveness may be defined, with reference to the sketch.

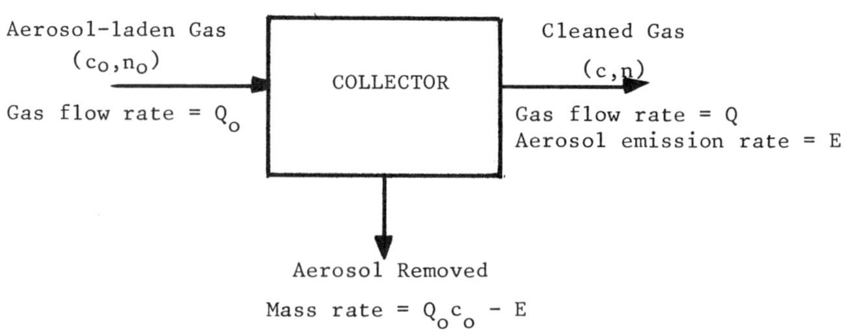

Let Q = volume rate of flow of gas stream e.g., m^3/min; cfm

c = mass of aerosol per unit volume of gas, called "grain-loading" e.g., gm/m^3;gr/ft^3 [a]

n = number of aerosol particles per unit volume of gas e.g., m^{-3}; ft^{-3}

E = mass emission rate of uncollected aerosol e.g., gm/min; lbm/hr

$E = Qc$

Subscript o refers to entering condition.

A. Overall Performance

The following "overall collection efficiencies" have been defined and are in use for different purposes. These definitions, as written, assume that $Q_o = Q$.

Collector Efficiency, by Mass of Particles:

$$\eta_M = \frac{c_o - c}{c_o} = \frac{Q_o c_o - E}{Q_o c_o} = 1 - \frac{E}{Q_o c_o} = \frac{C}{Q_o c_o} \tag{1}$$

Collector Efficiency, by Surface Area of Particles:

$$\eta_A = \frac{\text{Surface area of particles collected}}{\text{Total surface area of particles entering}} \tag{2}$$

Collector Efficiency, by Number of Particles:

$$\eta_N = \frac{n_o - n}{n_o} = \frac{Q_o n_o - Q_n}{Q_o n_o} \tag{3}$$

These three efficiencies depend upon d_p^3, d_p^2 and d_p respectively, where d_p represents particle size. They are, therefore, quite different numerically when applied to the same collection. The experimental

[a] A grain (gr.) is defined as 1/7000 or a pound mass.

determination of any of these quantities depends greatly upon the techniques available for measuring either number of particles, surface area, or mass.

The overall collector efficiency by mass of particles is usually the easiest to measure experimentally. The inlet and outlet gas streams in steady flow may be sampled by a collecting device, such as an absolute filter, which collects virtually all of the dust. By weighing the dust collected from these samples, and by measuring the volume of gas which simultaneously passed through the samples, the values of c_o and c may be calculated. Alternatively, only the inlet (or outlet) stream need be sampled, and the dust collection rate measured. In these calculations, if the temperature and/or pressure of the outlet stream differs much from that of the inlet, correction will have to be made for the fact that $Q_o \neq Q$.

Some on-stream monitoring devices actually function by counting particles, or by measuring their surface through an indirect effect such as light-scattering. In a case of this kind, either η_N or η_A is obtained by the test procedure. If a sample is taken for examination under the microscope, η_A will usually be based upon the projected area of particles.

Designing a collection system to minimize the overall dust emission rate, E, will usually be an important criterion of performance. This will be true especially whenever a system must meet legally specified emission limitations such as those discussed in Chapter 1. The following terms are defined to focus attention upon emission rather than collection.

Penetration, based upon mass emission:

$$P = E/Q_o c_o = c/c_o = 1 - \eta_M \tag{4}$$

(alternatively, may sometimes be based upon number of particles n/n_o)
Decontamination Factor, reciprocal of penetration:

$$D.F. = c_o/c = 1/P \tag{5}$$

(alternatively, may be taken as n_o/n)

$$\text{Decontamination Index} = \log_{10} (\text{D.F.}) = \log_{10} 1/P \qquad (6)$$

$$\text{Number of transfer Units NTU} = \ln 1/P = 2.30 \text{ D.I.} \qquad (7)$$

Equations (4), (5) or (6) are especially useful in expressing collector performance whenever there is a very small amount of uncollected dust. For example, it is more graphic to compare a performance of 99.0% mass collection efficiency with one of 99.9%, or 99.99% in terms of one of these items:

Overall mass efficiency	$(100\eta_M)$	:	99.00%	99.90%	99.99%
Penetration	(100 P)	:	1.0%	0.1%	0.01%
Decontamination Factor	D.F.	:	100	1000	10,000
Decontamination Index	D.I.	:	2	3	4
Number of Transfer Units	NTU	:	4.605	6.908	9.210

An improvement in performance by one unit in the Decontamination Index, represents reducing the dust emission rate to one-tenth of its previous value from the same incoming stream. The concept of a Transfer Unit has been borrowed from the N.T.U. used by chemical engineers to express the performance of a gas absorption tower or a distillation column. It is useful because several of the mathematical models of collector performance are exponential in nature, as will be shown later.

Example 1

An air stream containing 5 grains of dust per standard cubic foot is passed through a particle collector at the rate of 10,000 std. cu. ft. per min. (commonly abbreviated scfm). Samples of the aerosol are taken before and after the collector, and the number of particles of various sizes in an equal volume of air are counted (say by a technique such as the Coulter Counter). The results of the counts are tabulated below. They are presented in an over-simplified form in comparison with actual test results, in order to reduce complexities in calculation and in order to focus attention upon the basic principles.

Particle Size -Microns	Number/Volume n_o - Input	Number/Volume n -Output
100	100	1
50	900	27
10	1100	44
6	1500	255
2	1000	280
1	500	250

The particles may be regarded as spherical with the size given being the diameter, and as all having a density of 2.5 gm/cm^3.

(a) What is the overall efficiency of this collector as expressed by mass, by surface area, and by number of particles? (b) What is the penetration? The outlet dust loading? (c) What is the emission rate in lbm/hr?

Solution: it is assumed that the surface of each particle is proportional to the square of its diameter and the volume of each particle proportional to the cube of the diameter. Only relative values of surface and volume are required. A table may be constructed as shown below in which the relative surface proportional to nd_p^2, and the relative volume proportional to nd_p^3, of all particles of each size are tabulated. Calculations are then carried out as shown below the table, with references to the numbered equations. Note that particles of different sizes are collected with different degrees of effectiveness.

B. Grade or Fractional Efficiency

It might be imagined that, for a given kind of collection mechanism operating upon particles of a given kind, there would be a critical particle size such that all particles larger than this size would be collected completely, and all smaller particles not collected at all. However, every type of collecting force operates in a manner which depends upon and varies with particle size, as well as shape, and density. Consequently, different particle sizes are collected with different degrees of effectiveness, as indicated in the above example. The overall collection is a composite or summation of these actions.

The relationship between collection efficiency and particle size, for particles of a given kind being treated in a given collector is called the "grade-efficiency" or "fractional efficiency." It may be expressed either in tabular, graphical, or explicit functional form, $\eta_i = f(d_{p_i})$. Here d_{pi} represents the size of an individual particle of a certain "kind" as indicated by the subscript i. Kind refers to particles having a certain density, shape and composition. The grade-efficiency of i is the efficiency with which particles of a certain kind are collected, as a function of their size.

The grade-efficiency may be expressed in terms of any of the same measures as are used for overall efficiency. However, for particles of fixed size and all of the same "kind," the grade-efficiency by number of particles, by surface area of particles and by mass or volume of particles are all identical numbers.

For most collectors, operating upon a dust composed of particles all of the same kind, there is a unique single-valued function f for the particular set of operating conditions. This function will, however, depend upon such parameters as the nature and design dimensions of the collector, rate of flow and loading of gas stream, temperature, nature of collecting forces, etc. In the case of fabric filters ("baghouses") these statements must be modified, as will be shown in Chapter 8.

If the dust is a mixture of particles of more than one kind, there will be a different function of particle size for each kind. These functions will differ because of the effect of density and shape of particles upon the collection process. In what follows, it will be assumed usually that we are dealing with a single kind of particle, although whenever possible the effect of particle density will be explicitly stated.

Example 2.1

Particle Size d_p -microns	No./Vol. Input No.	No./Vol. Output No.	d_p^2	d_p^3	$n_o d_p^2$	nd_p^2	$n_o d_p^3$	$n \times d_p^3$
100	100	1	1×10^4	1×10^5	1×10^6	1×10^4	1×10^8	1×10^6
50	900	27	2.5×10^3	1.25×10^5	2.25×10^6	6.75×10^4	1.125×10^8	3.375×10^6
10	1100	44	1×10^2	1×10^3	1.10×10^5	4.4×10^3	1.10×10^6	4.4×10^4
6	1500	255	36	216	5.40×10^4	9180	3.24×10^5	55080
2	1000	280	4	8	4×10^3	1120	8×10^3	2240
1	500	250	1	1	500	250	500	250
Totals	5100	857	- - -	- - -	3.419×10^6	9.245×10^4	2.139×10^8	4.477×10^6

(a) Mass Efficiency $= \eta_M = \dfrac{\Sigma n_o d_p^3 - \Sigma n d_p^3}{\Sigma n_o d_p^3} \times 100$ (1)

$$= \frac{2.139 \times 10^8 - 4.477 \times 10^6}{2.139 \times 10^8} = 97.9\%$$

Surface Efficiency $= \eta_A = \dfrac{\Sigma n_o d_p^2 - \Sigma n d_p^2}{\Sigma n_o d_p^2} \times 100$ (2)

$$= \frac{3.419 \times 10^6 - 9.245 \times 10^4}{3.419 \times 10^6} \times 100 = 97.3\%$$

Number Efficiency $= \eta_N = \dfrac{\Sigma n_o - \Sigma n}{\Sigma n_o} \times 100$ (3)

$$= \frac{5100 - 857}{5100} \times 100 = 83.2\%$$

(b) Penetration $= P = 1 - \eta_M = (1 - 0.979) \times 100 = 2.1\%$ (4)

Outlet dust grain loading $= c = Pc_o = 0.021 \times 5$

$= 0.105$ grains/std. cu. ft.

(c) Emisson Rate $= E = PQ_o c_o = 0.021 \times 1{,}000 \times \dfrac{5 \times 60}{7000}$ (4)

$= 0.157$ lbm/hr

A study of experimentally determined grade-efficiency curves for various kinds of collecting devices, reveals that each kind of device tends to have a typical performance curve. The curves of Figures 2 and 3 are a selection of graphical representations of typical performance associated with certain standard types of equipment. Qualitatively they show, for example, that cyclones are generally less efficient than other kinds of equipment, that venturi scrubbers are more effective on finer particles than other scrubbers, that there may be a certain particle size for which collection is at a minimum, and other features. Each such graph is valid only for a fixed set of operating conditions for the collector it represents,

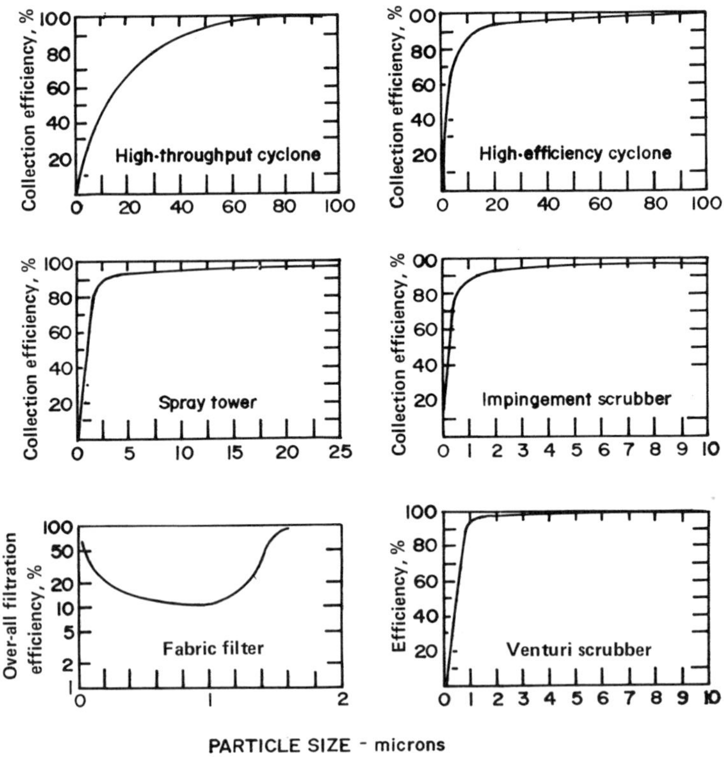

FIG. 2. Grade efficiency curves for various types of dust collecting, equipment [1]. Reproduced by permission from Chemical Engineering.

all of which must be stated in order for the graph to have quantitative meaning. Fig. 3 in particular shows the results of some recent experimental measurements in which special attention was directed toward the collection of very fine particles.

For purposes of illustration, the graph shown in Fig. 4 was selected as an example of a cyclone collector operating upon a speci-

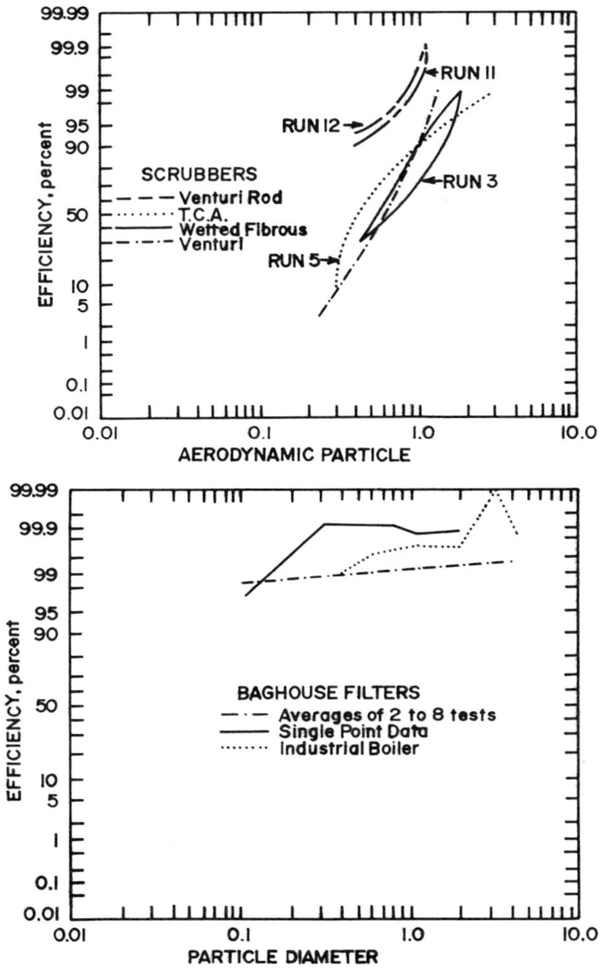

FIG. 3. Experimental measurements of grade efficiency [2]. Reproduced by permission from Chemical Engineering Progress.

fic dust of constant density. It is one of the goals of theoretical modelling of collector action to be able to predict a grade-efficiency function relationship in the form of an equation. For example, the "lateral-mixing" model (see page 95) for a cyclone collector leads to an equation of the form:

$$\eta_i = 1 - \exp - 2(A\, d_{p_i}^2)^{\frac{1}{2n+2}} \tag{8}$$

where A and n are given explicitly by the model theory in terms of design and operating parameters. This is discussed in detail in Chapter 6. This kind of equation does correspond to a graph like that in Figure 4.

An important special point on the grade-efficiency function is that at which η_i = 0.50. The particle size at this point is called the "cut size" or "cut diameter." This value will be referred to frequently in discussing collector performance.

C. Relation Between Grade Efficiency and Overall Efficiency

When the grade-efficiency relationship is known it may be used to predict the overall collection efficiency of an aerosol of a given

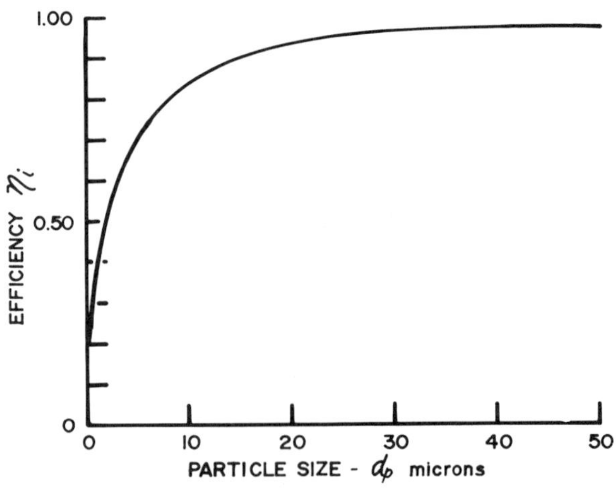

FIG. 4. Typical performance of cyclone collector. Design of cyclone, operating conditions, and kind of dust are fixed.

size distribution. There are two ways of doing so, depending upon whether the data are provided by grade-intervals, or by continuous distribution functions. In making these calculations, it is customary to assume that the dust is uniformly distributed in any inlet or outlet stream.

1. Grade-Interval Calculations. Given the grade- or size-distribution make-up of an aerosol, say in terms of mass fraction, g_i, of particles of size and kind designated by d_{p_i}: $gi = g_i(d_{p_i})$, where $\Sigma\, g_i = 1$, as discussed in detail in Chapter 2. Given also the grade-efficiency performance of the collector which is to be used to treat an aerosol such as described above, $\eta_i = \eta_i\,(d_{p_i})$.

The incoming aerosol mass rate of flow may be calculated for each size range as

$$\text{mass rate of input in } i^{th} \text{ grade} = Q_o c_o g_i \tag{9}$$

and emission rate for each size range as

$$\begin{aligned}\text{mass rate of emission in } i^{th} \text{ grade} &= (1 - \eta_i)\, g_i Q_o c_o \\ &= P_i g_i Q_o c_o\end{aligned} \tag{10}$$

Overall emission rate will be the summation of the grade rates, over all values of i:

$$E = Q_o c_o \sum_i (1 - \eta_i) g_i = Q_o c_o \left(1 - \sum_i \eta_i g_i\right) = Qc \tag{11}$$

Overall penetration will be given by (4):

$$P = E/A_o c_o = 1 - \sum_i \eta_i g_i = \sum_i P_i g_i \tag{4}$$

Overall collector efficiency by mass will be given by (1):

$$\eta_M = \frac{Q_o c_o - E}{Q_o c_o} = - P = \sum_i \eta_i g_i \tag{1}$$

The calculation procedure is illustrated by the following example.

Example 2. Consider the dust of size-distribution as given in Example 1 as being collected according to the grade-efficiency given in Fig. 4. (a) Calculate the penetration of the collector, and (b) determine the outlet dust size distribution.

Solution: (a) the procedure may be displayed conveniently in tabular form with the calculations represented by each column associated with a numbered equation above.

Particle Size d_{p_i} -microns	No/Vol Input- n_{o_i}	(A) $n_{o_i} d_{p_i}^3$	(B) g_{o_i}	(C) η_i	(D) $\eta_i g_{o_i}$
100	100	1×10^8	0.4675	100	0.467
50	900	1.125×10^8	0.5260	0.97	0.510
10	1100	1.10×10^6	0.0051	0.84	0.0043
6	1500	3.24×10^5	0.0015	0.73	0.0011
2	1000	8×10^3	3.7×10^{-5}	0.50	1.9×10^{-5}
1	500	500	2.3×10^{-6}	0.40	9.2×10^{-7}
TOTALS	5100	2.139×10^8	1.000	- -	0.982

The columns are determined as follows: (A) values are repeated from Example 1, (B) $g_{o_i} = n_o d_{p_i}^3 / \sum n_o d_{p_i}^3$ = mass fraction of particles of size d_{p_i}, (C) grade-efficiency read from Fig. 4, (D) product of entries in columns (B) and (C). Using Equation (4):

$$\text{Penetration} = P = 1 - \Sigma\, \eta_i m_i = 1 - 0.982 = 0.0176 \text{ or } \underline{\underline{1.76\%}}$$

It is clear from column (D) that the overall performance (penetration) of this collector is determined almost entirely by the grade-performance on the two coarsest size particles in the dust, for which the penetration is 1-(0.467 + 0.510) = 0.023, or 2.3%. However, if it should become necessary to reduce the total penetration say to below 0.5%, considerably improved collections of the finer sizes would be required.

(b) To determine the size distribution of the uncollected dust, it is simply necessary to calculate the uncollected portion of each size range as shown in this tabulation:

Particle Size-d_p -microns	Output $(1-\eta_i)n_i$	Input	Output
		Number fraction - f_i	
100	0	0.020	0
50	27	0.176	0.019
10	176	0.216	0.125
6	405	0.294	0.287
2	500	0.196	0.355
1	300	0.098	0.213
TOTALS	1408	1.000	1.000

A similar calculation could be done in terms of weight fractions.

2. Continuous Distribution Calculations

Given the size distribution in terms of a cumulative fraction-less-than-size d_p, $G_o = G_o(d_p)$, or $F_o = F_o(d_p)$ such as one of the functions described in Chapter 2. Given also the grade efficiency performance of the collector by an equation such as (8), $\eta = \eta\,(d_p)$. Then the overall efficiency, in terms of mass, will be given by

$$\eta_M = \int_0^1 \eta dG_o = \int_0^\infty \eta q_o \, dd_p \tag{12}$$

The overall penetration will be given by

$$P = 1 - \eta_M = \int_0^1 (1 - \eta) dG_p = \int_0^\infty (1 - \eta) q_o dd_p \tag{13}$$

The overall emission rate will be given by (4). The size distribution of the emission particulate, by cumulative fraction-less-than size, $G = G(d_p)$, may be determined by a material balance taken over all particles finer than a given size d_{p_i}:

$$(A_o c_o) G_{o_i} = G_i \cdot E + G_{c_i} \cdot C \tag{14}$$

and

$$Q_o c_o \int_0^{G_{o_i}} \eta_i dG_o = CG_{c_i} \tag{15}$$

Combining

$$G_i = \frac{1}{P}\left[G_{o_i} - \int_0^{G_{o_i}} \eta_i dG_o\right] \tag{16}$$

Evaluation of the integrals in (12), (15) and (16), may be done either analytically or graphically. If equations are given explicitly in terms of d_p for both the η-function and the G-function, it may be possible to work out the integral rigorously. Otherwise, graphical or numerical approximate integrations will be required.

a. Analytical Evaluations. A fairly common case will be that where G (and q) is a log-normal distribution (Equations (21), (26), (27), of Chapter 2) and η is an exponential function like (8), viz

$$\eta = 1 - \exp - Md_p^N \tag{17}$$

This is a Weibull type function (Equation (36), Chapter 2). Where these relationships hold (13) may be displayed as

$$P = \frac{1}{\sqrt{2\pi}} \int_{\infty}^{\infty} e^{\frac{-t^2}{2} Md_{Pm}^{-N} \sigma_g^{Nt}} dt \tag{18}$$

where $t = [\ell n\ d_p/\bar{d}_{p_m}]/\ell n\sigma_g$. While this cannot be integrated analytically, charts could be prepared by numerical integration, for given sets of values of $\bar{d}_{p_m}$ (MMD), σ_g, M and N. Chapter 6 gives such a chart for cyclones.

Another case, studied by Sundberg [3], is that in which both G_o and η may be represented by log-normal functions. Sundberg shows empirically that grade-efficiency test data for some collectors may be represented approximately by straight lines on the special log-probability graph paper described in Chapter 2. Then the "cut diameter" d_{p_c} (at which $\eta = 0.50$) and a standard geometric deviation for grade-efficiency σ_{g_c} are the parameters for η, just as $\bar{d}_{p_m}$ and σ_g are those for G. Sundberg proves that (12) yields:

$$\eta_M = \text{erf}\left[\frac{\ln\ \bar{d}_{p_m}/d_{p_c}}{\sqrt{\ln^2\sigma_g + \ln^2\sigma_{g_o}}}\right] \tag{19}$$

and gives a pair of charts by which this calculation may be made. It could also be done on a pocket calculator using the methods mentioned in Chapter 2 to find the value of erf.

Vatavuk [4] assumed that particle size distribution might sometimes be represented by an exponential function also, at least over the major part ($0.15 < G < 0.85$) of the range:

$$G_o = 1 - e^{-\beta d_p}$$

This is a special case of the Rosin-Rammler distribution, equation (45) Chapter 2, in which s = 1. Coupling this with (17), equation (13) becomes

$$P = \beta\int_0^\infty \exp - (Md_p^{\ N} + \beta d_p)\ dd_p \tag{20}$$

In general this must be evaluated numerically, but when N = 1 (a special case of some interest) this result is simply

$$P = \beta/(M + \beta) \text{ or } \eta_M = M/(M + \beta) \tag{21}$$

Equation (16) will yield

$$G = 1 - \exp - (M + \beta) d_p \tag{22}$$

b'. Graphical Methods. The integral in (12) may be evaluated by making a plot of η vs. G on which each point corresponds to a certain particle size. This is shown pictorially in Figure 5. The area under the entire curve is η_M. This value will be located at a horizontal line such that the shaded areas are equal.

The integral in (16) is represented by the cross-hatched area under the curve up to any desired value of G_o. This will have to

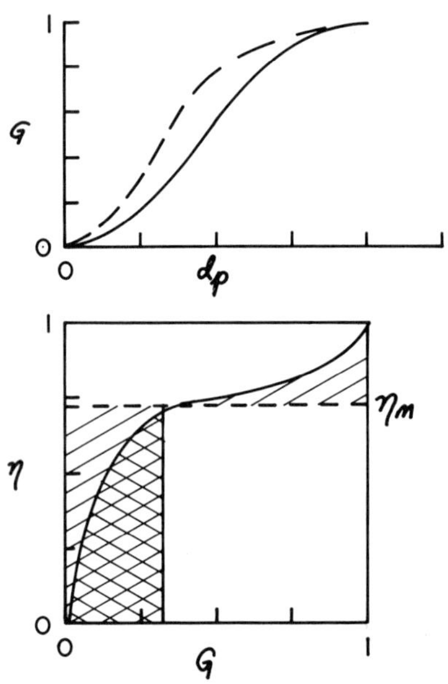

FIG. 5. Graphical determination of overall efficiency (12) and outlet particle size distribution (16).

be determined for a number of values of G_o in order to construct the outlet size distribution. The value of this integral, subtracted from G_o at each particle size, is used in (16) in order to locate points on the outlet curve. This is also shown on Fig. 5. The Gauss method of quadrature has been found to be an effective method of numerical integration in these cases. It lends itself well to pocket calculator computations.

Numerical illustrations will be given later along with the discussions of individual collector performance.

D. Interrelationships Among Particle-Size Functions

It is now evident that there are three important and interrelated functions of particle size: (i) the size-distribution of particles per unit volume in the input stream; (ii) the grade-efficiency relationship; (iii) the size-distribution of particles per unit volume, in the output stream. Only two of these are independent; if any two of them are given the third one is fixed and may be calculated. The three possible cases are: (1) to predict the output (iii), from a specified collection process (ii) acting upon a specified input dust (i)-a collector design procedure, as illustrated in Example 2 above. (2) To determine the grade-efficiency performance (ii) by obtaining a size-distribution analysis of both the input (i) and output (ii) streams-a collector performance testing procedure, discussed further and illustrated below. This is sometimes referred to as a "reverse grade-efficiency" procedure. (3) To determine an input size-distribution (i) by obtaining an output-size-distribution (iii) from a known grade-efficiency performance (ii)-a method of particle size analysis, the reverse of the calculations in Example 2.

1. Collector Design

The most important tool needed in collector design calculations is the ability to predict the grade-efficiency function for a given collector operating under specified conditions. Consequently, the central aspect of the theory of collection is the development of mathematical models which will show exactly how grade-efficiency is dependent upon operating parameters such as flow rates, grain-loading, dust properties, collector shape and dimensions, etc. With a complete model in hand an engineer may predict the output to be obtained from a given input. He may test the effect of alternative designs, and the effect of varying the operating parameters in order to arrive at an optimum design. Chapters 6, 7, 8 and 9 are devoted to the development and application of such models.

It happens that because the lateral-mixing model, discussed on page 95, is frequently used, the resulting grade-efficiency function is of the form of (8) or (17)

$$\eta = 1 - \exp - M d_p^{N}$$

Here N (usually $0 < N \leq 2$) is given by the model theory and M is a (known) function of the collector design and operating conditions. A common problem is to find a value of M to meet a required overall efficiency. Thus, for (1), (4) or (12), (13)

$$P = 1 - \eta_M = \sum_i g_{o_i} \exp - M d_{p_i}^{N} \tag{23}$$

is to be solved for M, given P, N, and a set of g_{o_i}. The input size distribution data may be arranged so as to select values of d_{p_i} such that all g_{o_i}'s are equal, say $g_{o_i} = g = 0.1$. Letting $x = e^{-M}$, (18) becomes

$$P = g \sum x^{d_{p_i}^N} = g(x^{d_p^N 1} + x^{d_p^N 2} + \ldots) \tag{24}$$

Because of the increasing values of d_{p_i}, the higher power terms in this series become small very rapidly, and using only the first term:

$$x \approx (P/g)^{1/d^N_{p_i}} \text{ or } M \approx -\ln(P/g)/d^N_{p_i} \tag{25}$$

If necessary, two or three terms of (24) may be used in a trial and error solution for M, using (25) as a very good first approximation. Illustrations of this method will be given in later chapters.

2. Collector Testing

It is necessary to carry out experimental measurements of dust size distribution and grain-loading of both the input and the output streams in order to establish the real performance of collectors, and incidentally to provide data for checking theoretical models. Because of the experimental difficulties in making particle size-distribution measurements upon flowing gas streams, many published tests of collector performance report only overall efficiencies. There is a very limited amount of data available which is sufficiently complete to be useful in testing of theoretical models. It is not within the purpose of this book to discuss the details of experimental techniques, but it is important to state what type of data is needed and the principles involved in its use.

The determination of a mass grade-efficiency relation requires the measurement of the mass of aerosol per unit volume of gas, of each size-grade in both the input and the output streams, that is c_{o_i} and c_i. Then

$$\eta_i = 1 - c_i/c_{o_i} \text{ or } P_i = c_i/c_{o_i} \tag{26}$$

However, instruments for performing a size-distribution analysis may only give the percentage of each size present, i.e. g_{o_i} and g_i. In this case, the total mass of all particles per unit volume of gas ("grain-loading") must also be measured. Then

$$\eta_i = 1 - \frac{c\, g_i}{c_o g_{o_i}} \text{ or } P_i = \frac{c}{c_o}\,\frac{g_i}{g_{o_i}} \tag{27}$$

Alternatively, the total dust flow rates in and out may be used for $c/c_o = E/Qc_o = P$ from (14) and

$$\eta_i = 1 - P_i = 1 - P\,\frac{g_i}{g_{o_i}} \tag{28}$$

It is to be noted that measurement of g_i and g_{o_i} only is insufficient to determine the grade-efficiency function. Example 3 illustrates the use of experimental test data to calculate a grade-efficiency curve, the so called "reverse grade-efficiency" method.

Example 3. Data for the particulate emissions from a coal-fired power plant upstream and downstream from an electrostatic precipitator have been published by Holland and Conway [5]. From their plot of cumulative mass distribution, values equivalent to g_{o_i} and g_i have been determined and are tabulated below.

i	Range of d_{p_i} -microns	Mass-fraction in range: g_{o_i}-up stream	Mass-fraction in range: g_i-down stream	Calculated: $P\,\frac{g_i}{g_{o_i}}$	Calculated: η_i
1	0.6	(0.020)	(0.070)	- -	- -
2	0.6-0.7	0.004	0.01	0.05	0.95
3	0.7-0.8	0.004	0.02	0.10	0.90
4	0.8-1.0	0.007	0.03	0.086	0.914
5	1-2	0.035	0.14	0.08	0.92
6	2-3	0.060	0.16	0.053	0.947
7	3-4	0.240	0.29	0.024	0.976
8	4-5	0.130	0.01	0.002	0.998
9	5-6	0.020	0.0	0	1
10	6-8	0.020	0.02	0.02	0.98
11	8-10	0.030	0.025	0.017	0.983
12	10-20	0.110	0.085	0.015	0.985
13	20-30	0.080	0.070	0.018	0.982
14	30	(0.24)	(0.07)	- -	- -

The overall mass efficiency of the precipitator was not reported, but for purposes of illustration here may be reasonably assumed, in line with Fig. 2, to be 98%, i.e. P = 0.02. Determine the grade-efficiency function.

Using (23), values of P_i and η_i may be calculated and tabulated, as in the last two columns. The values of g_{o_i} and g_i and η_i are plotted in Figure 6. Care must be used in discussing these graphs because the raw data were taken from a small-scale plot which is difficult to read precisely. However, a few comments may be appropriate. Since the dust is the fly-ash from a coal-fired boiler, it is undoubtedly a mixture of several kinds of particles. This is further indicated by the presence of two predominant particle size-ranges, one of 3-4 microns, the other between 10-20 microns. It is not surprising, therefore, to find that the grade-efficiency curve has a rather irregular appearance, although this could be due in part to the uncertainties in the raw data. The value of grade-efficiency in any one particle size-range probably represents a composite of the collection action of the precipitator upon the different kinds of particles present.

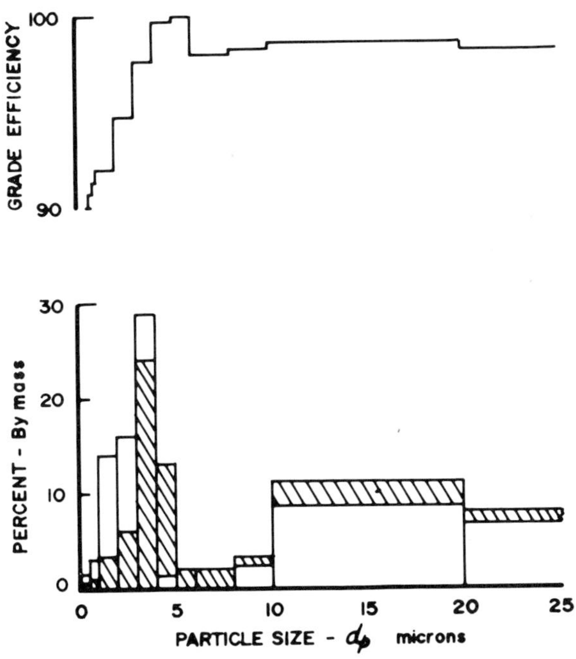

FIG. 6. Experimental performance of electrostatic precipitator, Example 3.

If the overall efficiency of the precipitator is other than the assumed value of 98%, the grade-efficiency graph will be displaced accordingly higher or lower, but its general trend will not be affected.

Occasionally the particle-size-distribution of the collected dust may be of interest, and in fact may be easier to obtain from a sample than in the case of the flowing gas streams. From equation (14) G_{c_i} may be found from G_{o_i}, G_i and P.

$$G_{c_i} = 1\ [G_{o_i} - (1-\eta_M)\ G_i]/\eta_M$$

where G_{c_i} represents the cumulative per-cent-less-than size d_{p_i} in the collected dust. If the overall collector performance is known, any one of these three size distributions in (14) may be determined from the other two, even though the grade-efficiency function is not specifically known.

In principle G could also be used in (15) as another way of determining the grade-efficiency

$$\int_0^{G_{o_i}} \eta_i dG_{o_i} = (1-P)\ G_{c_i}$$

This might be used to infer η from data giving G_o and G_{c_i} as functions of d_{p_i}. This would be done by taking successive intervals of particle size $(d_{p_{i+1}} - d_{p_i})$, and using the corresponding intervals of ΔG_{o_i} and ΔG_{c_i}, calculate η_i for the midpoint of the interval

$$\eta_i \simeq (1-P)\ \Delta G_{c_i}/\Delta G_{o_i} = (1-P)\ g_{c_i}/g_{o_i}$$

III. BASIC MODELLING CONCEPTS

A. Residence Time of Particles

A fundamental factor which influences whether a given aerosol particle will be collected is the length of time during which that particle comes under the influence of the collecting force(s). This time must be sufficient for the particle to travel at the collecting velocity from its point of entrance into the zone of influence of the collecting force, to a point of collision, before the particle is swept out of the zone of influence by the air stream velocity.

The length of time a particle is within the zone of influence is called its residence time, t_r. For collision to occur, the residence time t_r must be equal to or greater than the collecting time. The collecting time t may be defined as:

$$t = \frac{\text{distance travelled from point-of-entry to collecting surface}}{\text{collecting velocity}} \tag{29}$$

The sketches in Figure 7 illustrate how the relationship between t_r and t may be affected by : (a) location of point of entry; (b) difference in air stream velocity; and (c) difference in particle size, or mass.

For the purpose of this elementary discussion, the collecting velocity is regarded as constant. In fact, however, when a particle enters a collecting zone, it undergoes an initial period of acceleration in the direction of the collecting force. In many instances, this period is very short and for practical purposes, the particle may be regarded to have attained a steady collecting velocity almost instantaneously.

An overall average residence time for all particles in a stream may be calculated from the free volume of the collector and the volume flow rate.

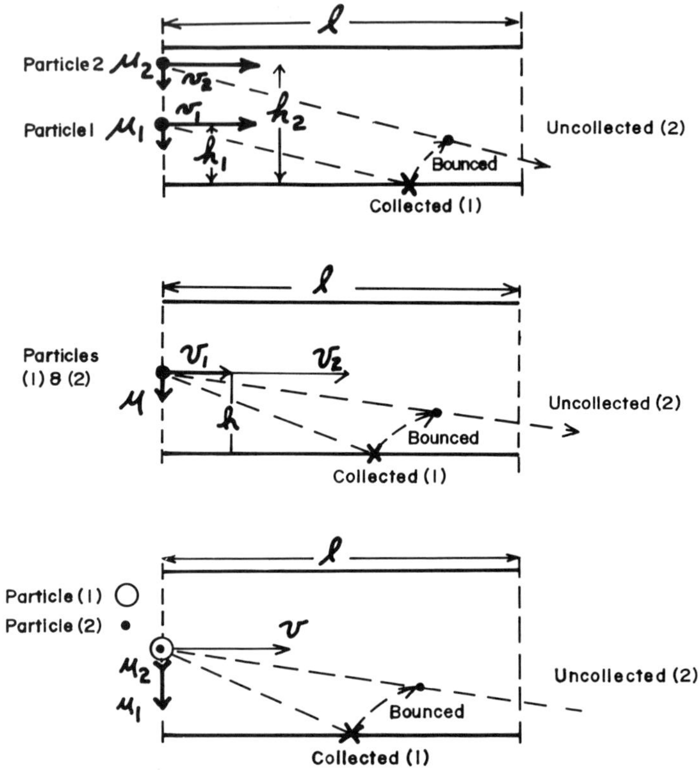

FIG. 7. Relation between collecting time and residence time. (a) Effect of location of point of entry; (b) effect of air stream velocity; (c) effect of particle size or mass.

$$t_{r_{avg}} = \frac{\text{volume of zone of influence in collector}}{\text{volumetric flow rate of air stream}} = \frac{V}{Q} \tag{30}$$

The actual residence time of an individual particle may be quite different from this value, depending upon the nature of the fluid velocity profile.

(a) In the case of plug, or piston flow, all parts of the air stream are moving with the same velocity v, and the residence time for all particles is the same:

$$t_r = \frac{V}{Q} = \frac{\ell}{v} \tag{31}$$

where ℓ = length of zone of collection, in direction of flow.
(b) In the case of a real laminar or turbulent flow, there will be a fluid velocity distribution pattern over the cross-section of the collection zone perpendicular to the direction of flow. The precise description of this pattern must be known, in order to calculate the residence times for particles entering along different streamlines.

Sometimes a particle may bounce off after a collision with a collecting surface. It will then be carried farther along by the air stream, at the same or a different air stream velocity depending upon the flow pattern. It may even be carried out of the collecting zone, as shown in Fig. 7, which assumes plug flow for the sake of simplicity in the illustration. The particle which is uncollected due to the bounce effect may be regarded as equivalent to one: (a) entering at a larger collecting distance $h > h$; or (b) having a lower collecting velocity, $u < u$; or (c) traveling with a higher air stream velocity, $v > v$, as indicated by the lines marked "Bounced" in Fig. 7. In each case, the effect is to give the particle an apparent collecting time which is greater than its residence time.

A particle may also rest on, or move along, the collecting surface for a time and then become re-entrained due to some mechanical action (such as "rapping" the plate of an electrostatic precipitator) or by a current of higher-velocity air. The collecting surface does, in time, have to be cleaned of collected particles in order that it may continue to function, or else it must be renewed in some way. If the aerosol particles are liquid, or the surface is flushed with a liquid, or coated with some oily or adhesive substance, then the effects of bouncing or re-entrainment are greatly reduced or eliminated.

Another effect which must be considered is that which eddy currents in the fluid may have upon suspended particles. Such currents will tend to keep the uncollected particles mixed back into the air stream, and prevent them from moving steadily toward the collecting surface at a theoretical collecting velocity. Instead, their paths would consist of erratic wanderings back and forth due to eddy currents, or in the case of very small particles due to their bombardment by gas molecules (Brownian diffusion). There would, however, be a net motion toward the collecting surface, because on the average, the eddies would carry the particle in either direction at random with equal probability. The zone of influence of the collecting force is really then only fully effective in the immediate vicinity or the collecting surface, and in the bulk of the nominal collecting zone the particles may be more or less well-mixed.

B. Elementary Models

There are three idealized situations which may be identified as basic models to serve as a starting point to describe real collection processes: (1) Plug flow, no radial or axial mixing of uncollected particles; (2) complete radial (lateral) mixing of uncollected particles; (3) complete back-mixing, both radially and axially.

Each of these models may be used as the basis for predicting a corresponding form of grade-efficiency relationship. The nature of this form will also depend to a very significant extent upon the way in which the collecting velocity u_i may depend upon d_{p_i} in the particular application of the model. The possible relationships between u_i and d_{p_i} are presented in the detailed discussion of collecting forces given in Chapters 4 and 5. The three models will be described in general form at this point for application to specific collectors later.

1. Plug Flow, No Mixing

This model may be illustrated in terms of a rectangular two-dimensional view of the zone of influence. It is assumed that the particles are distributed uniformly across the section, that all particles are of the same size and kind, d_{p_i}, that the collecting force is operating only in one direction, and that the trajectories of all particles comprise a family of parallel straight lines. Whether a

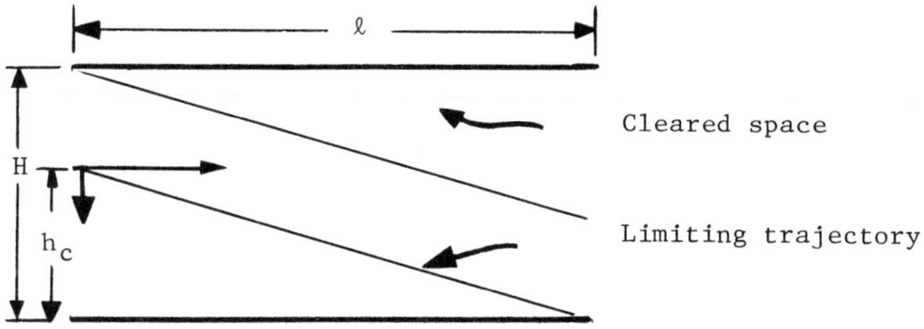

given particle is collected then is determined solely by the distance h from its point of entry to the collecting surface. All those particles for which this distance $h < h_c$ will collide with the surface. The shaded space in the sketch becomes completely cleared of particles. The efficiency of collection, by number or by mass, will then be given by

$$\eta_i = h_{c_i}/H = u_i\ \ell/vH \tag{32}$$

The grade-efficiency relationship will be determined by the way in which the velocities u_i depend upon d_{p_i}. If $u_i \propto d_{p_i}^2$, which is a common case, then η_i vs. d_{p_i} will be a parabolic graph with $\eta_i = 1.00$ for all particles large enough so that $u_i \geqq \frac{vH}{\ell}$ (as in Figure 8).

2. Lateral Mixing

This model assumes that the uncollected particles are well-mixed by lateral turbulence in the gas stream, hence their concentration is always uniform across any lateral or radial section perpendicular to

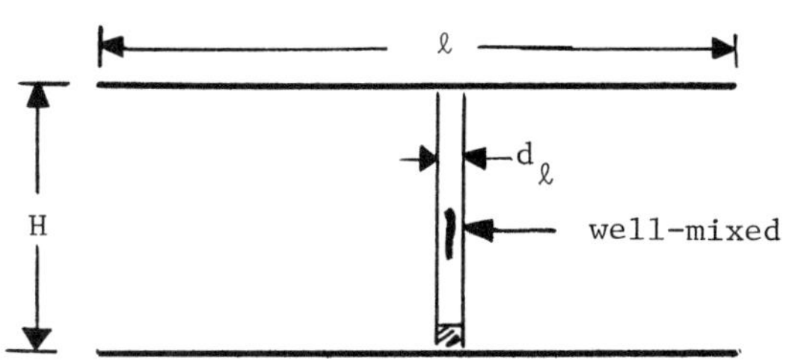

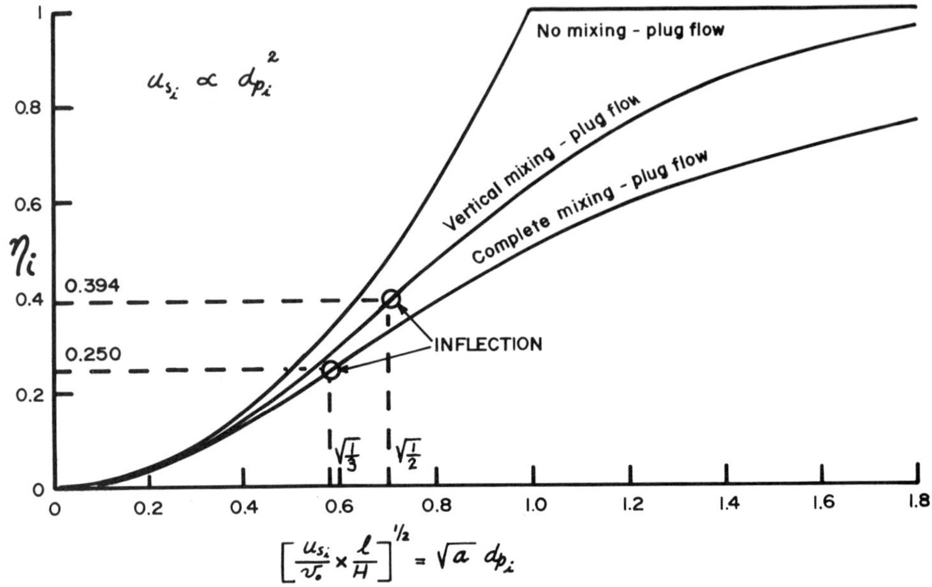

FIG. 8. Generalized grade efficiency for gravity collector.

the collecting surface. Plug flow is assumed in the axial direction. As the stream flows over a differential distance $d\ell$ during a period of time dt, all particles of the ith grade within a distance $u_i dt$ of of the collecting surface will move into that surface. There will occur a fractional reduction in particle concentration which is the same everywhere along the direction of flow. It is given by:

$$-\frac{dn_i}{n_i} = \frac{u_i dt}{H} = \frac{u_i}{H_v} d\ell$$

This may be integrated over the length of the collecting zone to give:

$$-\int_{n_{i_o}}^{n_{i_o}} \frac{dn_i}{n_i} = \frac{u_i}{vH} \int_o^{\ell} d\ell = \frac{u_i}{v}\frac{\ell}{H}$$

$$\ln n_i/n_o = -u_i \ell/vH = \ln P = -NTU \quad (33)$$

The collection efficiency is then:

$$\eta_i = 1 - \exp(-u_i \ell/vH) = 1 - \exp(-u_i t_r/H) \tag{34}$$

where $t_r = \ell/v$ is the residence time. Again, the grade-efficiency relationship will depend upon how u_i is related to d_{p_i}, but in any event the exponential function will require that η_i approach 1.00 asymptotically as d_{p_i} increases. If $u_i \alpha d_{p_i}^N$, the equation becomes

$$\eta_i = 1 - \exp(-M d_{p_i}^N) \tag{35}$$

where M represents all the factors which do not depend upon d_{p_i}. This is the most commonly used model as has been mentioned in (17) above.

If N = 2 this expression possesses a point of inflection at

$$d_{p_{inf}} \sqrt{1/2M} \text{ and } \eta_{inf} = 0.294 \tag{36}$$

The graph will, therefore, be S-shaped as shown in Fig. 8.

3. Back Mixing

This model assumes that the degree of turbulence in the gas is such that at any instant all uncollected particles are uniformly mixed throughout the entire volume of the collector. During a unit period of time, per unit width of collecting space, dust collected = $(u_i \ell n)$ particles/time; dust leaving, uncollected = $(v_o H\ n)$ particles/time. The grade-efficiency is given by

$$\eta_i = \frac{u_i \ell\ n}{u_i \ell\ n + v_o H n} = \frac{u_i \ell / v_o H}{1 + u_i \ell / v_o H} \tag{37}$$

This functional relationship also has η_i approaching 1.00 asymptotically as d_{p_i} increases. For the case of $u_i \alpha d_{p_i}^2$, it becomes

$$\eta_i = \frac{M\, d_{p_i}^2}{1+Md_{p_i}^2} \tag{38}$$

which has a point of inflection at

$$d_{p_{inf}} = \sqrt{1/3M} \text{ and } \eta_{inf} = 0.250 \tag{39}$$

and is, therefore, also S-shaped when graphed.

A comparison of all three models is shown in Fig. 8 for the case of a gravity collector where the collecting velocity is proportional to $d_{p_i}^2$.

It should be clear at this point why the grade-efficiency relationship never corresponds to a precise cut-off at a certain critical particle size. Any real collector operation is going to have effects present such as are embodied in each of these models. Either the effect of the position of particle at entry, or the effect of back mixing will be such as to require a gradual change in efficiency with particle size, rather than a step jump from 0% to 100% at a certain size.

4. Application of Models

The basic models of collection efficiency just described may be applied to the performance of specific collecting devices in either of two ways: by the study of the fundamentals of particle motion, or by the empirical analysis of experimental data. The details of the fundamental approach form the subject matter of the later chapters which deal with each type of collector individually.

It is evident that the exact nature of the grade-efficiency relationship corresponding to each model will depend upon the manner in which the collecting velocity is related to particle size. It involves an identification of all the forces acting upon a particle, and the use of Newton's law of motion to predict the path taken by the particle under the combined action of these forces. To apply this information to a given kind of collector requires a detailed knowledge of the nature of the collecting forces operating in that collector. This, in turn, leads to the functional relation between u_i and d_{p_i}. The fundamentals of particle mechanics are presented in Chapters 4 and 5.

These studies show that, in many cases, $u_i \propto d_{p_i}^2$. This corresponds to a range of conditions in which the motion of the particle is governed by what is called Stokes Law. So common is this situation, that it was used above as the basis for the graphical illustration of the models. However, other relationships such as $u_i \propto d_{p_i}$ are possible. A more general treatment of the models can be made by taking

$$u_i \propto d_{p_i}^N$$

where N may range from -1 to +2 in value. There is a range of conditions just outside of those corresponding to Stokes Law in which the value of N may lie between 1.6 and 2.0.

With different values of N, the appearance of the grade-efficiency graphs will change markedly. For example, the model graphs may not exhibit a point of inflection, nor a range of downward concavity. As an aid to identifying possible relationships, general formulas have been derived for the value of η at the point of inflection as follows:

Lateral-mix model: $$\eta_{inflec} = 1 - \exp(1 - N)/N \tag{40}$$

Back-mix model: $$\eta_{inflec} = (N - 1)/2N \tag{41}$$

Typical values are tabulated below. Since real values of η must lie between 0 and 1, a negative value of η_{inflec} means that no point of inflection will appear.

N	Grade-Efficiency at Inflection Lateral-Mix	Back-Mix
-1	Negative	1
0.5	Negative	Negative
1	0	0
1.5	0.283	0.167
1.6	0.313	0.188
1.7	0.338	0.206
1.8	0.359	0.222
1.9	0.377	0.237
2.0	0.393	0.250

It is noteworthy that values of η_{inflec} are at the low end of the range.

Furthermore, there is no requirement that the value of N be the same throughout the entire range of operating conditions for a given type of collector. If the nature of action of collecting forces changes with particle size, or with other parameters of operation, N may shift in value from one end of the range to the other. This can give rise to what might be called a "composite model." Such models are observed in practice, for example, the graph for a filter in Fig. 2. Without the insight obtained from the principles of particle mechanics, such behavior would be difficult to understand.

5. Testing of Models

The other approach to the application of the models involves the empirical examination of an experimentally determined grade-efficiency curve. Simple graphical tests may be made to determine whether one of the models is applicable, and if so, the value of N.

a. To test for model (1) - plug flow, no mixing: Plot $\log \eta_i$ vs. $\log d_{p_i}$. If the model fits, a straight line will be obtained with slope equal to N. For if $u_i = a\, d_{p_i}^N$; from (27)

$$\log \eta_i = n \log d_{p_i} + \log \left(\frac{a\ell}{vH}\right) \tag{42}$$

b. To test for model (2) - lateral-mixing only:
Plot $\log [-\ln(1-\eta_i)]$ or log NTU vs. $\log d_{p_i}$. If the model fits, a straight line will be obtained with slope equal to N. For, if $u_i \alpha d_{p_i}^N$, from (15)

$$1 - \eta_i = \exp\left(-\frac{u_i l}{vH}\right) = \exp\left(-Md_{p_i}^N\right)$$

$$-\ln(1 - \eta_i) = NTU = Md_{p_i}^N$$

$$\log[-\ln(1 - \eta_i)] = \log NTU = \log M + N\log d_{p_i} \qquad (43)$$

c. To test for model (3) - complete back-mixing:
Plot $\log \eta_i/(1 - \eta_i)$ vs. $\log d_{p_i}$. If the model fits a straight line will be obtained with slope equal to N. For if $u_i \propto d_{p_i}^N$, from (38)

$$\eta_i = \frac{Md_{p_i}^N}{1 + Md_{p_i}^N}$$

$$\frac{\eta_i}{1-\eta_i} = \frac{Md_{p_i}^N}{1 + Md_{p_i}^N} \frac{1}{1 + Md_{p_i}^N} = Md_p^N$$

$$\log \frac{\eta_i}{1-\eta_i} = \log M + N \log d_{p_i} \qquad (44)$$

From a glance at the typical performance graphs presented in Fig. 2 and 3, it might appear that none of the models could fit most of these curves because the curves exhibit neither a point of inflection nor a range where they are concave downward. There are several possible explanations to account for this:

(1) The experimental data in the low end of the particle-size range may be either inaccurate or incomplete to the extent that a point of inflection is not revealed even though it may exist.

That is, the curve as drawn may simply be roughly extrapolated down to $d_{p_i} \to 0$ from the lowest data point, say in the range of $d_{p_i} = 1.5\ \mu m$. Experimental measurements on very fine particles are very difficult to make accurately.

(2) One of the models may fit with a value of $N \neq 2$, such that it does not call for an S-shaped curve.

(3) Due to particle bouncing or re-entrainment, the collection efficiency over part of the range of particle sizes may be less than called for by the model.

(4) The value of N, or indeed the nature of the applicable model itself may shift over the range of particle size.

Example 4. Test the graph presented in Fig. 4 for agreement with each of the basic models.

In order to test all three models it is necessary to plot η, $\ln 1/(1 - \eta)$, and $\eta/(1 - \eta)$ respectively, against d_p on log-log paper, as given by equations (42), (43) and (44). This may be done conveniently by first preparing a table of these values based upon values of η_i read from Fig. 4 for selected values of d_{p_i}. This table is as follows.

d_{p_i}-μm	η_i	$\ln \frac{1}{1-\eta_i}$	$\frac{\eta_i}{1-\eta_i}$
2	0.50	0.693	1
4	0.65	1.05	1.86
6	0.73	1.31	2.70
10	0.84	1.83	5.25
20	0.93	2.66	13.3
30	0.965	3.35	27.6
40	0.975	3.69	39.0

All three graphs may be made on the same chart by using 2 x 2 cycle log-log paper and defining the ordinate scale appropriately for each function of η. These plots are given as Figure 9.

The plot of $\log \ell n\, 1/(1-\eta_i)$ vs. $\log d_{pi}$ gives a good straight line, while the other two plots are curved, thus confirming agreement with the lateral mixing model. The value of N, equal to the slope of this line, may be calculated by selecting two points, say for $d_{p_i} = 2$ and 30 μm:

$$N = \text{slope} = \frac{\log 3.35 - \log 0.693}{\log 30 - \log 2} = 0.582$$

IV. ENERGY REQUIREMENTS AND EFFICIENCY

The energy required to operate a dust collection system will be mainly of two kinds: (a) work required to move the dust-laden air stream through the equipment; (b) additional energy of a form unique to the type of collection force being employed. The first kind is

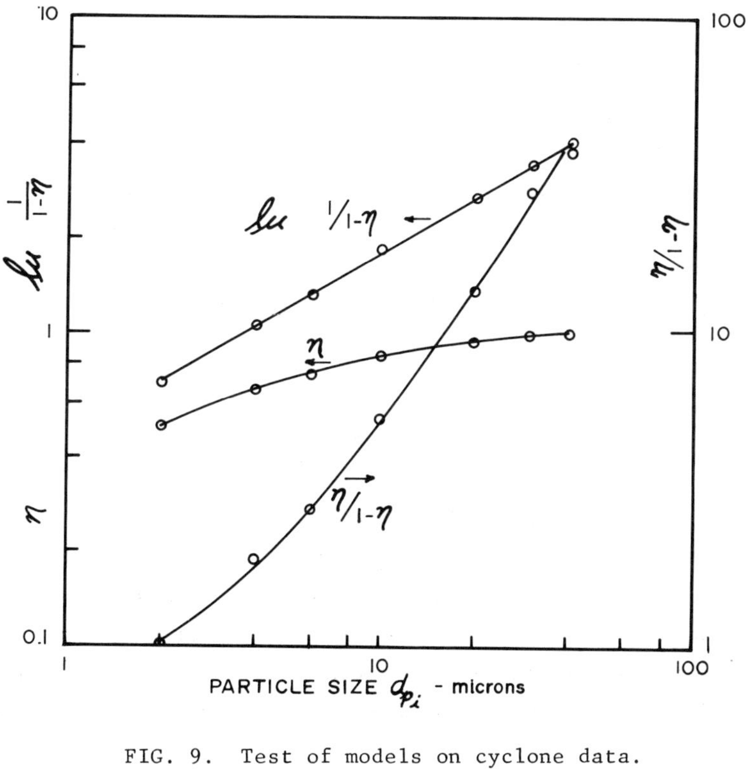

FIG. 9. Test of models on cyclone data.

due to the friction losses incurred by the flow of the stream through the equipment. The second kind may be electrical, as in an electrostatic precipitator; heat, as in a thermal depositor; other mechanical work, as in pumping a scrubbing fluid, or in shaking a fabric filter to clean it of collected dust.

The second kind should never be overlooked. However it may be completely absent in certain mechanical collectors, such as cyclones, and even when present may in many cases be very small relative to the first kind.

The first kind is always present. Hence it is always important to be able to estimate the pressure drop in the air stream, as it flows through the system, and to calculate the corresponding power loss which must be provided by the fan or blower which moves the stream. Thus,

$$\text{Power loss} = Q \times \Delta P/33{,}000 \quad h_p \tag{45}$$

where Q = flow rate, in cubic feet per minute, and ΔP = total pressure drop, in lbf per square foot.

Since Q may be very large, it is important to strive for low ΔP, which is then usually expressed in inches of water. Values of 10 inches water or less are rather common, although in certain types of collectors ΔP may range upwards to ten times this value. For these units:

$$\text{Power loss} = 1.58 \times 10^{-4}\ Q \times \Delta P''\ \text{hp.} = 1.18 \times 10^{-4}\ Q\ \Delta P''\ \text{kw} \tag{46}$$

where $\Delta P''$ = inches of water. A stream of 10,000 cfm at 10 inches water will thus require 15.8 hp or 11.8 kw.

The total pressure drop due to frictional losses will be made up of losses in the duct work plus loss in the collector itself. The first may be calculated by the standard methods of fluid mechanics, in which it is usually assumed that the air behaves in a noncompressible manner. The latter requires special treatment which may be different for each type of collector. Thus, along with the modelling for collector efficiency, it is equally important to develop a model for pressure drop, i.e. frictional loss, for each collector. Both must be considered in the design.

The scheme which is used to estimate frictional energy losses is based upon a mechanical energy balance across the collecting device.

$$\Delta \left(\frac{v^2}{2g_c}\right) + \frac{\Delta P}{\rho_f} + g\,\frac{\Delta Z}{g_c} = W_s - F \tag{47}$$

Each term represents energy per unit mass of fluid stream. The delta (Δ) terms are taken as the difference between a point in the exit duct immediately beyond the collector and a point in the inlet duct just upstream of the collector. If these ducts have approximately the same cross-sectional area, $\Delta(v^2) \simeq 0$, and if they are at the same, or nearly the same, elevation $\Delta Z \simeq 0$. Furthermore there is usually no work exchanged between the gas stream and its surroundings between these two points so that $W_s = 0$. The equation usually reduces to $\Delta P/\rho_f = -F$.

It is then convenient to evaluate F as a multiplier of the inlet velocity head $v_o^2/2gc$, say $F = N_H\ v_o^2/2g_c$ where N_H is called the "number of inlet velocity heads" frictional loss. Finally, the pressure drop is expressed as

$$\Delta P = -\ \rho_f N\ v_o^2/2g_c \tag{48}$$

Most applications of energy loss calculations are thus resolved into a method of predicting N_H for a given collector operating under a given set of conditions.

In a wet-scrubber type of collector, there is also the energy required to pump the scrubbing fluid. This is calculated in a manner similar to equation (44) except that the units are usually different. Expressing ΔP in psi (lbf/in^2) and Q in gpm (gallons/min), the calculation becomes:

$$\text{power loss} = 5.83 \times 10^{-4}\ Q\ \Delta P\ \text{hp} = 4.35 \times 10^{-4}\ Q\ \Delta P\ \text{kw} \qquad (49)$$

If some liquid is lost in the process, this may be expressed as equivalent to a power consumption:

$$\text{Equivalent power loss} = 60\ Q_c\ (\frac{\$}{\text{gal}}) \times (\frac{\text{kw-hr elec.}}{\$})\ \text{kw} \qquad (50)$$

where Q_c = water consumed, gpm; \$/gal = cost of scrubbing liquid; and \$/kw-hr = cost of electricity. The total power consumption for the scrubber would then be given by the sum of (46), (49) and (50).

Some typical power consumption values for common dust collecting equipment are given by Sargent [1], Stairmand [6], Vandergrift [7], and Benson [8]. They range from 0.2 kw/1000 cfm to 4 kw/1000 cfm for various kinds of equipment, capacities and collection efficiency. In every case it is found that the higher the overall collection effiency, the higher the power requirement. As a very rough rule of thumb, an average value may be taken as 1 kw/1000 cfm.

Example 5. Estimate the total electrical power required to process 1000 acfm of gas in each of the following cases, assuming the energy efficiency of pumps and fans to be 75%: (a) cyclone having 6-in water pressure drop; (b) Venturi scrubber having 20-in water pressure drop in the gas, 20 psi pressure drop for 7 gpm scrubbing water, with 4% loss of water at \$1.05/1000 gal, electricity at \$0.03/kw-hr; (c) electrostatic precipitator having 0.5-in water pressure drop and using 0.3 kw of electricity.

Solution: with reference to equations given above:

(a) Equation (46)

$$\text{power consumed by fan} = \frac{1.18 \times 10^{-4} \times 1000 \times 6}{0.75} = \underline{\underline{0.94}}\ \text{kw/1000 cfm}$$

(b) Fan-power (Equation 46)

$$= \frac{1.18 \times 10^{-4} \times 1000 \times 20}{0.75} = 3.15\ \text{kw/1000 cfn}$$

Pump power (Equation 49)

$$= \frac{4.35 \times 10^{-4} \times 17 \times 20}{0.75} = 0.08 \text{ kw/1000 cfn}$$

Water loss (Equation 50)

$$= 60 \times 7 \times \frac{0.04}{1000} \times \frac{1.05}{0.03} = 0.59 \text{ kw/1000 cfn}$$

Total equivalent power consumed = 3.15 + 0.08 + 0.59 = 3.82 kw/1000 cfm

(c) Fan power (Equation 46)

$$= \frac{1.18 \times 10^{-4} \times 1000 \times 0.5}{0.75} = 0.08 \text{ kw/1000 cfm}$$

Electrical power for collection = 0.3 kw/1000 cfm

Total power consumed = 0.08 + 0.3 = 0.38 kw/1000 cfm

The energy efficiency of the collection process is an important consideration to which little attention has been given. Stukel and Rigo [9] have taken as an approximation to the minimum amount of work required to separate the particles from the gas, simply the kinetic energy of the particles removed. Per unit time:

$$W_{min} = \eta_M (Q_o c_o) v^2/2 \tag{51}$$

For typical conditions of $\eta_M = 0.95$, $c_o = 5$ gr/ft^3, v = 50 ft/s this will give

$$W_{min} = \frac{0.95}{32.2} \times \frac{1000}{60} \times \frac{5}{7000} \times \frac{(50)^2}{2} \times \frac{0.745}{550} = 5.95 \times 10^{-4} \text{ kw/1000 cfm}$$

This corresponds to an energy efficiency of about 0.059%. The entire range of cases considered by Stukel and Rigo fall within energy efficiency values of 0.008 to 0.1%. Similar calculations by Soo [10] also show the efficiency of energy utilization in common collectors to be well below 0.5%. There is a need to learn ways of separating particulates from gases with a much improved energy efficiency. Stukel and Rigo remark that these efficiencies are "intolerably low."

V. GENERAL COMPARISON OF COLLECTORS

It is helpful, particularly in making a preliminary survey of choice of collecting equipment, to have some general method of overall comparison of performance. Several such methods are available and may be used as rough guides.

A convenient method of comparing the grade-efficiency performance of various collectors is to replot data such as given by Fig. 2 as a semi-log plot of log $(1-\eta_i)$ vs. d_{p_i}. As an illustration of this, a few selected cases taken from Fig. 2 are shown in Fig. 10.

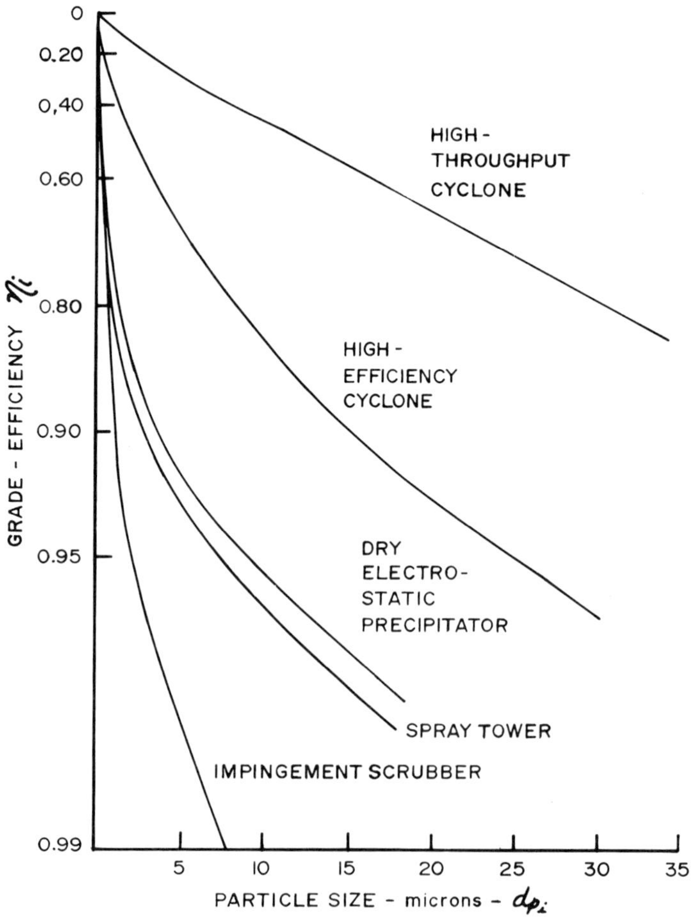

FIG. 10. Comparison of collector performance.

This method of plotting gives all curves the common point in the upper left corner where $d_{p_i} = 0$ and $\eta = 0$ or $P = 1.00$. It is really a plot of log P vs d_p. The less efficient performance is represented by a curve with a smaller slope i.e. extending farther to the right. For the cases shown, performance could be ranked in descending order, as

1 - High throughput cyclone
2 - High efficiency cyclone (Fig. 3)
3 - Dry electrostatic precipitator
4 - Spray tower
5 - Imprringement Scrubber

which is very readily observed by a glance at the chart.

One should be careful however only to draw very rough general conclusions from such comparisons. It must be remembered that the precise performance curve for any kind of collector depends upon operating parameters. The curves shown are examples only. They could each be shifted to some extent in either direction by a variation in gas flow rate, nature of dust, temperature, or other conditions. One cannot say, for instance, that all scrubbing collectors are more efficient (or less efficient) than all electrostatic precipitators.

There is no reason to expect that such plots should be straight lines. It may easily be shown that a straight line would correspond to a special case of the lateral-mix model with N = 1, i.e. $u_i \propto d_{p_i}$. For from equations (35)

$$\ln (1 - \eta_i) = \ln P_i = - Md_{p_i} \tag{52}$$

or $2.303 \log (1 - \eta_i) = - Md_{p_i}$, wo that $- M/2.303$ is the slope of this line, always negative. Such a case as this rarely occurs. If $N \neq 1$, the lines will be curved.

Stairmand [11] has provided some very useful charts which take into account costs as well as efficiency. As a basis for his comparison of the performance of collectors he defined three typical dusts designated as "coarse," "fine," and "superfine." The size distributions of these are given in Table 1. He then selected typical

Table 1

Particle Size Distributions of Standard Dusts [11]

Particle Size microns (μm)	% by weight less than size		
	Superfine Dust	Fine Dust	Coarse Dust
150	-	100	-
104	-	97	-
75	100	90	46
60	99	80	40
40	97	65	32
30	96	55	27
20	95	45	21
10	90	30	12
7.5	85	26	9
5.0	75	20	6
2.5	56	12	3

grade-efficiency performance for each of 18 different types of collectors at three different particle sizes: 50 μm, 5 μm, 1 μm. The overall mass efficiency of collection by each collector was then calculated for each of the three types of dust. This made it possible to list the collectors in order of increasing overall efficiency on each kind of dust. The list is in Table 2.

Next an estimate of the cost of each kind of collecting equipment was made on the basis of a set of assumed conditions, which are stated below. The total cost, including capital charges, per $1000 m^3$ of gas treated was plotted against the overall collection efficiency on a log-probability chart as shown in Figure 11 for the fine dust, and on Figure 12 for all three dusts. It is extremely interesting to note that a straight line correlating total cost with overall efficiency is fairly well defined on this type of plot. Stairmand's charts as originally published express costs in pence as of the period of the late 1960's. Because of fluctuating relationships between British and U.S. currency, and because of increases due to inflation since that time, it is necessary to regard the cost numbers as relative rather than absolute figures.

Table 2
Efficiency of Various Gas-Cleaning Equipment on Three Typical Dusts [11]

Dust Collector	Collection Efficiencies (%) on		
	Coarse Dust*	Fine Industrial* Dust	Super-fine* Industrial Dust (Fume)
Inertial collector	81.2	58.6	14.3
Medium-efficiency cyclones	84.6	65.3	22.4
Low resistance cellular cyclones	91.3	74.1	29.9
High efficiency cyclone	93.9	84.2	52.3
Impingement scrubber (Doyle type)	95.7	87.9	60.8
Self-induced spray deduster	97.6	92.3	70.3
Void spray tower	97.9	94.5	83.7
Fluidised-bed scrubber	99.0	96.2	82.7
Irrigated-target scrubber (Peabody type)	99.3	97.9	91.8
Electrostatic precipitator	99.5	98.5	94.8
Irrigated electrostatic precipitator	99.6	99.0	97.4
Flooded-disc scrubber-low energy	99.85	99.5	98.3
Flooded-disc scrubber-medium energy	99.90	99.7	99.0
Venturi-scrubber-medium energy	99.94	99.8	99.3
High-efficiency electrostatic precipitator	99.96	99.85	99.4
Venturi-scrubber-high energy	99.97	99.90	99.6
Shaker-type fabric filter	99197	99.92	99.6
Reverse-jet fabric filter	99.98	99.95	99.8

*For size distribution see Table 1

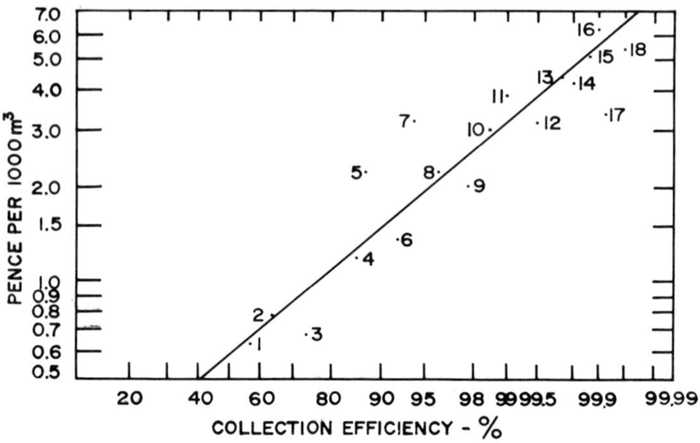

FIG. 11. Cost of cleaning equipment. Treating fine dust [11].

1. Inertial collector
2. Medium efficiency cyclone
3. Low resistance cyclone
4. High efficiency cyclone
5. Impingement scrubber
6. Self induced spray
7. Void spray tower
8. Fluidized bed scrubber
9. Irrigated target
10. Electrostatic precipitator
11. Irrigated electrostatic precipitator
12. Flooded disk-low energy
13. Flooded disk-medium energy
14. Venturi-medium energy
15. High efficiency ESP
16. Venturi-high energy
17. Shaker fabric filter
18. Reverse-jet fabric filter

Reproduced by permission from Filtration/Separation.

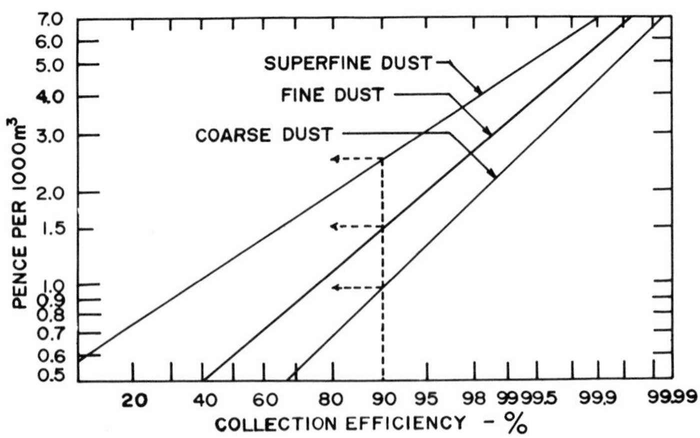

FIG. 12. Cost of cleaning equipment [11]. Reproduced by permission from Filtration/Separation.

Many caveats must be observed in using such charts, particularly those relating to all of the conditions assumed for the basis of comparison. In specific cases, deviations from one or more of the parameters assumed could make considerable difference in actual performance or cost. The assumptions made are tabulated, so that estimates of correction may be made for them when necessary. However, it must be emphasized that this method remains as a general overall type of comparison and is no substitute for specific detailed design calculations.

Assumptions made in comparison are:

Gas flow rate: 100,000 m^3/hr at 20°C (58,800 scfm)

Dust loading: 10 gm/m^3 (4.37 grains/ft^3)

Dust Size: as given in Table 1

Combined fan and motor efficiency: 60%

Annual operating time: 8000 hr.

Amortization rate: 10%

Capital cost includes auxiliaries such as fans, pumps, and motors, and allowance for site clearance and erection as follows: 100% of cost of equipment, except for electrostatic precipitators; 50% of cost of equipment for electrostatic precipitators.

In addition to this list of assumptions other factors must be kept in mind in referring to these graphs. They are based upon equipment available, and practices which are followed in Great Britain. There may be differences in details of design of certain items in the United States. This is another reason for regarding the cost figures to be purely relative, rather than absolute, especially when applied to U.S. conditions. Also, one must be very careful in reading the scales of Fig. 11 and 12, especially the probability scale used for the efficiency axis. Only a few tenths, or even hundredths, of percentage point lie across the spread of the performance of collectors numbered 12-18, i.e. those above 99.5%. A very small change in some operating condition could easily result in shifting the position of one collector in this group. One could not, for instance, make a sweeping generalization such as "A reverse jet fabric filter (No. 18) is always the most efficient collector" just by glancing at Fig. 11. Finally, it must be pointed out that there are other considerations which will enter into the choice of a collector for a given duty besides its efficiency and its cost although these are certainly very important.

The concepts discussed in this chapter may be summarized by a general system chart, Figure 13. This shows that the principal inputs to the collector system performance are: (a) the type of collector and its specific dimensions, (b) the operating conditions (gas flow, temperature, grain-loading etc.), and (c) the particle-size distribution of the particulate to be collected. The interaction of (a) and (b) produces the grade-efficiency which interacts with (c) to determine the outputs of the system: particulate emitted and particulate collected (quantities and size distribution). Finally (a) and (b) also determine the energy requirements, including the pressure drop.

The chapters in Part II have as their objective the application of this general system model to the specific collectors discussed. Each has a detailed system chart pertaining to the particular forces, design, and conditions involved.

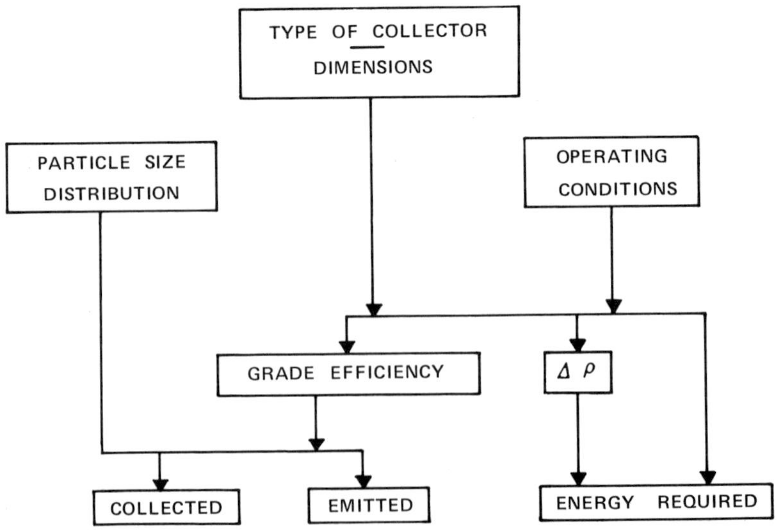

FIG. 13. General systems analysis of particualte collector performance.

REFERENCES

1. C.D. Sargent, *Chem. Eng.*, 130 (Jan. 27, 1969).

2. J.H. Abbott and D.C. Drehmel, *Chem. Eng. Prog.*, 72: No. 12, 47 (Dec. 1976).

3. R.E. Sundberg, *J.Air Poll. Cont. Assoc.*, 24: 758 (1974).

4. W.M. Vatavuk, EPA - 450/2 73-002, Aug. (1973).

5. W.D. Holland and R.E. Conway, *Chem. Eng. Prog.*, 69: 93 (1973).

6. C.J. Stairmand, *J. Inst. Fuels (London)*, 29: 58 (1956).

7. A.E. Vandergrift, L.J. Shannon, and P.G. Gorman, *Chem. Eng.*, 80: Deskbook Issue, 107 (June 18, 1973).

8. J.R. Benson and M. Corn, *J. Air Poll. Cont. Assoc.*, 24: 340 (1974).

9. J.J. Stukel and H.G. Rigo, *Atmos. Environ.*, 9: 529 (1975).

10. S.L. Soo, *Environ. Sci. and Tech.*, 7: 63 (1973).

11. C.J. Stairmand, *Filtration and Separation*, 7: 53 (1970).

CHAPTER 4

ELEMENTARY PARTICLE MECHANICS MOVEMENT OF AEROSOL PARTICLES IN STILL GAS

I. PARTICLE TRAJECTORIES

The study of particulate collection devices from a fundamental viewpoint requires a knowledge of the principles of mechanics which govern the motion of aerosol particles. The path followed by a particle while it is in the zone of influence of the collecting force(s) determines whether it will collide with the collecting surface or target. It is necessary to be able to predict this path, or trajectory, in order to develop models of collector performance. The methods of doing this will be presented in two parts: first, for motion in still air, and then for particles which are in a stream of moving air.

In certain types of collectors e.g., scrubbers and filters, there may also be a need to determine interaction between the fluid and individual target elements. For example, in a scrubber the target elements are drops of scrubbing liquid in motion relative not only to the dust particles, but also to the air stream. Certain aspects of target motion are treated by the same fundamental principles as apply to the determination of dust·particle trajectories. In what follows, wherever the word particle is used it may, in general, be understood to refer either to dust particles or to target elements.

A. Newton's Law

The motion of an individual particle, under the influence of various forces, is governed by Newton's Law of motion: Sum of all forces acting upon a particle = mass of particle x acceleration of particle. This is a vector differential equation: the various forces are vectors, and their sum is a vector addition. The resulting acceleration is a vector. In symbols it is usually written:

$$m_p \, d\bar{u}/dt = \Sigma \, \bar{F} \tag{1}$$

where the acceleration is represented as the derivative of the particle velocity vector.

A particle trajectory is determined by identifying all of the forces acting in a given situation and representing each of them by the appropriate function. Then the differential equation is solved (in principle) by integration in two stages: first to give the velocity as a function of time, and then, replacing $\bar{u}$ by ds/dt, to give distance (or position) as a function of time. This last equation is that of the trajectory of the particle. The solution is carried out by rewriting the vector equation in terms of its scalar component equations in one, two or three dimensions as may be required.

Each of the kinds of force which may be encountered in aerosol motion will be studied separately, and then in combination. Solutions will be presented for the various trajectories which are usually involved.

B. Forces on Particles

1. Drag Force

a. Drag Coefficients - Steady Motion. Whenever there is motion of a particle relative to the surrounding air, this motion will be resisted or opposed by a "drag force." This is the most fundamental force to be considered for it is always present, even though no other force may be acting appreciably. In particulate collector design work, there is a need to determine drag forces acting upon (a) dust particles themselves, and (b) target elements of target type collectors.

Drag arises out of two phenomena. Due to the shape of the particle (or target), fluid must be displaced around it as it moves. This sets up a greater pressure on the front of the particle than on the rear. The resulting force is called "form-drag." In addition, there is friction between the particle and the adjacent fluid, or between nearby layers of fluid. This results in "frictional drag" force. The two are usually considered together as producing one overall effect, or force. This is obviously determined by factors such as shape and size of particle, nature of particle surface, speed of motion, nature and properties of the fluid.

For a full treatment of the fluid mechanics of drag, the reader is referred to any standard text on fluid mechanics. In addition, there is an excellent little paperback by Shapiro called "Shape and Flow" [1]. This presents a well-illustrated account of experiments which show the fundamental phenomena involved, and the way in which drag forces depend upon the parameters noted.

The results of theoretical and experimental studies of drag lead to a fairly simple method of calculating it in any particular case. The total drag force may be represented in terms of a drag coefficient C_D, which is defined by this equation:

$$F_D = C_D \times A_{proj.} \times \rho_f u^2/2 \tag{2}$$

This is a scalar equation to give the magnitude of the vector drag force; the direction is exactly opposite to that of the velocity vector. In this equation the term $A_{proj.}$ represents the cross-sectional area of the particle (or target) image as projected in the direction of motion. The drag force per unit of projected area is seen to be proportional to the kinetic energy represented by a unit volume of fluid displaced at the particle velocity.

Values of the drag coefficient have been determined by applying Equation (2) indirectly to experimental data in which the equivalent of F_D, A_{proj}, and other terms have been measured. An orderly scheme has emerged for the correlation of these values with the other parameters. For particles of a given kind of shape, C_D may be correlated with the particle Reynold's number, defined as

$$Re_p = x_p \rho_f u/\mu = x_p u/\nu_f$$

where x_p is some appropriate linear dimension of the particle (or target), and ν_f is the kinematic viscosity of the fluid. Figure 1 shows this correlation for several shapes. It is apparent that there are several regimes of flow which may be identified by a corresponding range of values of Re. These data apply only to a particle in steady motion with a constant velocity relative to the fluid.

Although drag coefficients have been studied up to very large values of Re_p, as indicated in Fig. 1, applications to conditions where $Re_p > 800$ seldom arise in dust collection work. Further dis-

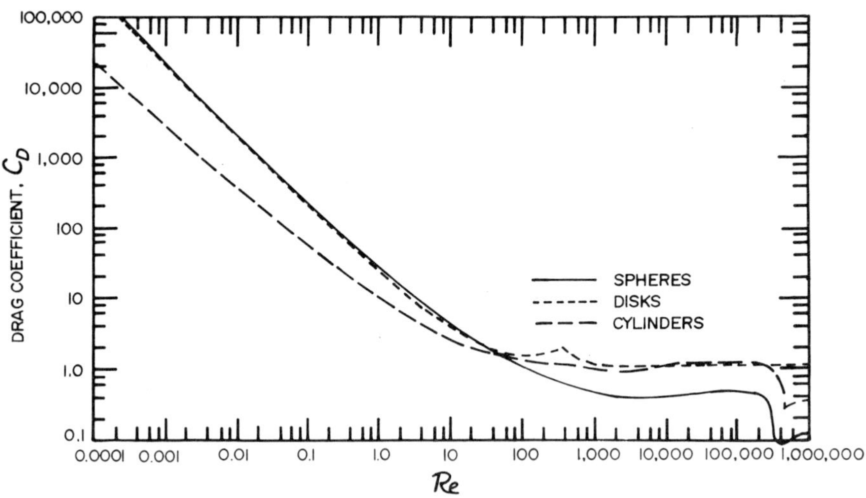

FIG. 1. Drag coefficients for spheres, disks, and cylinders [2]. Reproduced by permission from Perry's Chemical Engineers Handbook, 5th Edition.

cussion, therefore, will be limited to the lower range of Re_p values. This range may be subdivided into several portions according to the nature of the drag phenomena operative in each portion.

From this point on the particle will be regarded either as being spherical in shape with x_p replaced by d_p, the spherical diameter, or else as behaving in a manner equivalent to that of a spherical particle. A particle of irregular shape may thus be characterized by a "size" which is the diameter of an "equivalent" sphere.

Equivalent here is used in the sense of the motion of the sphere being the same as that of the particle. For material of unit density (1 gm/cm^3) the size is referred to as the "aerodynamic diameter" of the particle. In this context, the drag coefficient for spherical shape only may be examined in detail as shown in Figure 2. In the range of Re_p below approximately 0.1 there is evidently a simple relationship for C_D. This has been shown theoretically and confirmed experimentally to be

$$C_D = 24/Re_p \tag{3}$$

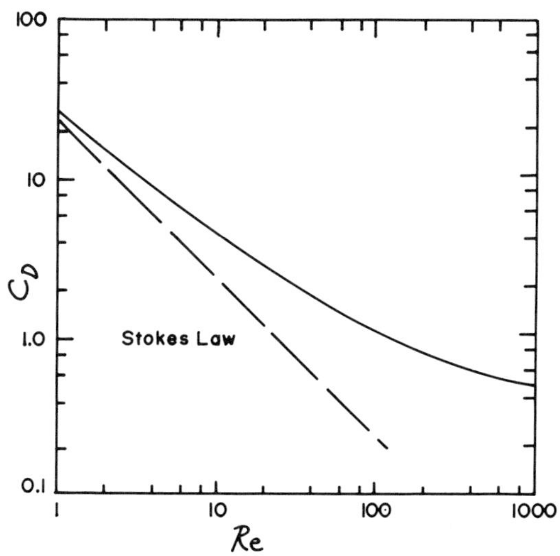

FIG. 2. Drag coefficients for spheres.

It corresponds to what is called Stokes law for drag. Thus

$$F_D = \frac{24}{Re_p} \times A_{proj} \times \frac{\rho_f u^2}{2}$$

For spherical particle $Re_p = d_p u/\nu$ and $A_{proj} = \pi d_p^2/4$. Then

$$F_D = \frac{24\mu}{d_p u \rho_f} \times \frac{\pi d_p^2}{4} \times \frac{\rho_f u^2}{2} = 3\pi d_p \mu u \qquad (4)$$

which is Stokes law [3]. It applies to a regime of pure laminar flow.

As the value of Re_p increases above 0.1, there is a gradual but definitely increasing deviation from the simple linear relationship of (3). The drag coefficient becomes greater than would be anticipated by Stokes law. This effect is attributed to the gradual onset of turbulence in the fluid surrounding the particle, and especially in the wake of the particle.

A number of equations, both semi-theoretical and empirical, have been proposed for the calculation of C_D from Re to give reasonable agreement with the experimental values. The following are recommended to be used up to $Re_p = 800$, in order to give agreement within ± 2% over this range, and at the same time to be relatively simple for calculation.

Range	Equation	Author	Ref.
$Re_p < 0.1$	$C_D = 24/Re_p$	Stokes	[3]
$0.1 < Re_p < 0.5$	$C_D = 24/Re_p + 4.5$	Oseen	[4]
$0.5 < Re_p < 3.0$ $400 < Re_p < 800$	$C_D = 24/Re_p + 3.60/Re_p^{0.313}$	Schiller & Naumann	[5]
$3.0 < Re_p < 400$	$C_D = 24/Re_p + 4/(Re_p)^{1/3}$	Klyachko	[6]

The Klyachko equation may be used over the entire range of Re_p from 0.5 to 800 with an error generally within ± 3-4%. It is superior to that of Dickinson and Marshall [7] which is stated to be valid within ± 7% for $Re_p < 3000$: $C_D = 0.22+24(1 + 0.15\ Re_p^{0.6})/Re_p$.

b. Applicability of Stokes Law: Cunningham Correction. In a great majority of applications to the motion of dust particles, such as are encountered in air pollution control problems, conditions fall within the realm of Stokes law. For example, consider motion in air at 20°C where $\nu_f = 0.151\ \text{cm}^2/\text{sec}$. In order for Re_p to be ≤ 0.1, it is necessary only that $d_p \leq 151/u$, d_p in microns, u in cm/sec. Figure 3 shows the maximum particle size which will be in the Stokes law range at various relative velocities and air temperatures.

So widespread is the applicability of this range that it is customary to assume Stokes law applies as a basic principle in much of dust collection theory. This greatly simplifies the calculation of drag forces. However, one should always check this assumption if errors of more than 2% arising from incorrect drag calculations are to be avoided. On occasion it may be necessary to use one of the more complicated methods of calculation of C_D according to the range of Re_p, instead of simply using Stokes Law. For example, this situation arises in calculating the motion of liquid droplets in scrubber

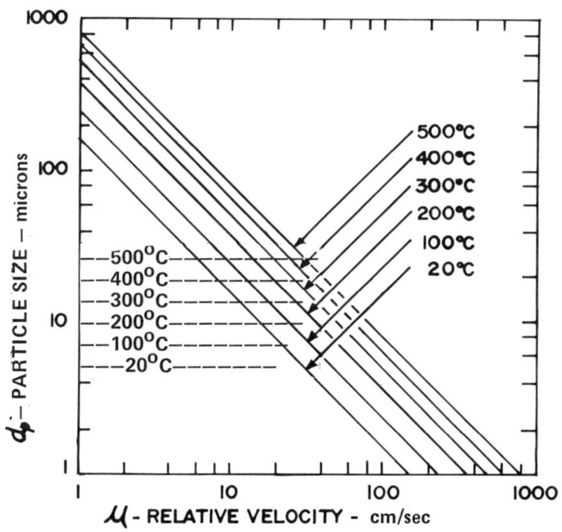

FIG. 3. Range of applicability for Stokes Law. Conditions of velocity, particle size, and temperature within triangular zone to left of slanting solid line and above dashed line.

operation, where values of d_p of the order to 200 microns and u of more than 6000 cm/sec are not uncommon.

Another correction must be taken into account whenever the particle size is small enough to be of the order of the mean free path (λ) of the gas molecules. Under this condition, the gas no longer behaves as a continuous medium with respect to the particle and its drag effect is reduced. To correct for this so-called "slip" condition, a correction factor known to the Cunningham factor C, is introduced into Stokes Law.

$$F_D = \frac{3\pi\mu d_p u}{C} = 3\pi\left(\frac{d_p}{C}\right)\mu u \qquad C > 1 \tag{5}$$

The value of C depends upon the ratio of $2\lambda/d_p$, called the Knudsen Number, Kn. C may be estimated from an equation due to Davies [8]

$$C = 1 + \frac{2\lambda}{d_p}\left[1.257 + 0.400 \exp\left(-1.10 \frac{d_p}{2\lambda}\right)\right] \tag{6}$$

which indicates that $C < 1.02$ when $Kn \leq 0.016$.

To use this equation Kn is calculated by taking λ = mean free path of gas molecules, based upon the Chapman-Enskog equation.

$$\lambda = \nu/0.499\bar{u} \tag{6a}$$

in which $\bar{u}$ = mean molecular velocity, cm/sec, given according to the kinetic theory by

$$\bar{u} = \sqrt{8RT/\pi M} \tag{6b}$$

from which, for a perfect gas,

$$Kn = 0.0226 \frac{\mu}{d_p P}\sqrt{\frac{T}{M}} \tag{6c}$$

where μ = gas viscosity, poises; T = absolute temperature °K; P = gas pressure, atmos.; d_p = particle diameter, cm.; M = molecular weight; R = gas constant, 8.31×10^7 ergs/°K mole.

This correction depends upon temperature, pressure, and particle size, being greater at higher temperatures, lower pressures, and for smaller particles. Figure 4 indicates the magnitude of the Cunningham correction to the drag force F_D by giving the value of d_p/C as a function of d_p at several temperatures, for dry air under 1 atmosphere pressure.

As a rough guide to the magnitude of the correction at 20°C, 1 atm: $C = 1 + 0.165/d_p$ (d_p in microns). For particles 8 microns and smaller the correction will be more than 2%. Fig. 3 indicates, at each temperature, the minimum particle size for which the correction will be less than 2% at atmosphere pressure. On this graph the triangular zone above and to the left of the lines marked with a given temperature indicates the range of velocity and particle size within which Stokes law may be used without correction, at that temperature.

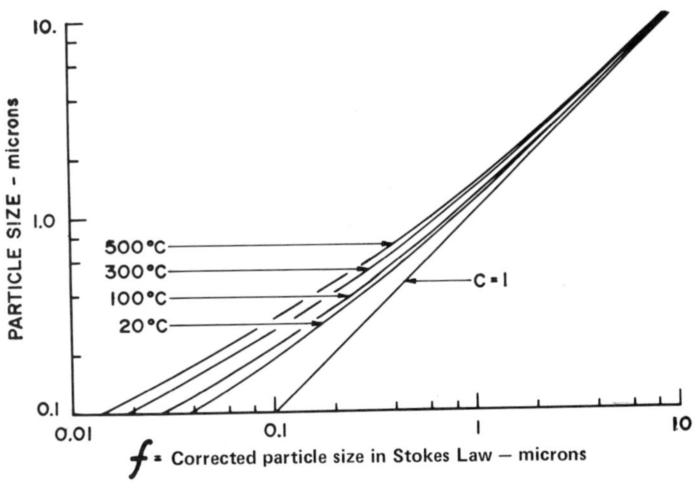

FIG. 4. Slip factor correction on Stokes law. Dry air at 1 atmosphere; $F_D = 3\pi f \mu u$; $C = d_p/f$ = Cunningham factor.

The validity of Equation (6) rests upon rather limited experimental data obtained only for air at temperatures up to 80°C and at pressures of 1 atmosphere and below. To apply it to other gases at elevated temperatures and pressures, must involve an extrapolation of an unknown degree of uncertainty. However, it is obvious that the mean free path should tend to be smaller at higher pressures and lower temperatures.

Willeke [25] has considered this problem from a theoretical point of view and derived formulas to be used at elevated temperatures. They are:

$$Kn = \frac{6.052 \times 10^{-4}}{d_p P} \quad \frac{T}{(1 + 110/T)} \tag{6d}$$

$$C = 1 + Kn\,[1.246 + 0.42 \exp - 0.87/Kn] \tag{6e}$$

in which T = °K, d_p = microns, P = atmospheres. They are probably better than (6) and (6c) but no data are available for checking.

The concept of a "Stokes Law Particle" is sometimes used to refer to an equivalent spherical particle moving under conditions such that Stokes law applies and the Cunningham factor is equal to 1, within a combined accuracy of say less than ± 1%. A whole set of special equations for various calculations may be developed for Stokes law particles. These will be indicated as they arise.

A more general definition of "aerodynamic diameter" (see page 121), incorporating the Cunningham factor, has been given by Calvert [19] as

$$d_{pa} = d_p\sqrt{C\rho_p}$$

This will be utilized especially in scrubber calculations, Chapter 9.

The procedure for determining an appropriate value for C_D and F_D is illustrated in the following example.

Example 1. Determine the drag force on a spherical particle moving through still, dry air in each of the following cases:

(a) d_p = 100 μm, u = 100 cm/sec, t = 20°C, P = 1 atm.
(b) d_p = 1 μm, u = 10 cm/sec, t = 100°C, P = 1 atm.
(c) d_p = 40 μm, u = 30 cm/sec, t = 20°C, P = 1/2 atm.
(d) d_p = 0.1 m, u = 10 cm/sec, t = 300°C, P = 1/2 atm.

For each case the procedure will be the same. First, calculate the Re_p. If $Re_p > 0.1$, Stokes Law will not apply and C_D must be calculated from an appropriate expression on page 122, or read from Fig. 2. The drag force will then be calculated from Equation (2). If $Re_p \leq 0.1$, Stokes law applies and the drag force may be calculated directly from Equation (4) provided no Cunningham correction is needed. To determine this, calculate Kn from Equation (6c) and if Kn > 0.016, determine C from (6). Then use (5) for the drag force. Whenever the pressure is equal to 1 atmosphere, Figures 3 and 4 may be used to shorten the work.

(a) At 20°C, 1 atm., $\mu = 1.81 \times 10^{-4}$ poises, $\rho_f = 1.205 \times 10^{-3}$ gm/cm^3, $\nu = 0.151$ cm^2/sec. Fig. 3 indicates no Cunningham correction is needed.

$$Re_p = \frac{100 \times 100^{-4} \times 100}{0.151} = 6.62 >> 0.1$$

Using Klyachko formula $C_D = 5.75$ (Fig. 2 reads $C_D = 5.6$)

$$F_D = 5.75 \times \frac{\pi(100 \times 10^{-4})^2}{4} \times \frac{1.205 \times 10^{-3}(100)^2}{2}$$
$$= 2.72 \times 10^{-3} \text{ dynes}$$

(b) At 100°C, 1 atm, $\mu = 2.17 \times 10^{-4}$ poises, $\rho_f = 0.940 \times 10^{-3}$ gm/cm^3, $\nu = 0.231$ cm^2/sec. Fig. 3 indicates that Stokes Law applies, but a Cunningham correction is needed. Fig. 4 reads f = 0.82 μm (C = 1/082 = 1.22)for d_p = 1 μm at 100°C

$$F_D = 3\pi\ f\ \mu\ u = 3\pi \times 0.82 \times 10^{-4} \times 2.17 \times 10^{-4}$$

$$= 1.68 \times 10^{-6} \text{ dynes}$$

(c) At 20 C, 1/2 atm. $\mu = 1.81 \times 10^{-4}$ poises, $\rho_f = 0.603 \times 10^{-3}$ gm/cm^3, $\nu = 0.302$ cm^2/sec. Since P ≠ 1 atm. Fig. 3 and 4 cannot be used.

$$Re_p = \frac{40 \times 10^{-4} \times 30}{0.302} = 0.397 > 0.1, \text{ Stokes Law does not apply.}$$

Using Oseen formula:

$$C_D = \frac{24}{0.397} + 4.5 = 65. \text{ (off-scale Fig. 2)}$$

For a particle of 40 μm, a Cunningham correction is probably unnecessary:

$$Kn = \frac{0.0226 \times 1.81 \times 10^{-4}}{40 \times 10^{-4} \times 0.5}\sqrt{\frac{293}{28.9}} = 6.5 \times 10^{-3} \quad C = 1.008$$

$$F_D = 65 \times \frac{\pi(40 \times 10^{-4})^2}{4} \times \frac{0.603 \times 10^{-3}(30)^2}{2} = 2.22 \times 10^{-4} \text{ dynes}$$

(d) At 300 C, 1/2 atm, $\mu = 2.93 \times 10^{-4}$ poises, $\rho_f = 0.307 \times 10^{-3}$ gm/cm^3, $\nu = 0.954$ cm^2/sec. Since P ≠ 1 atm. Fig. 3 and 4 cannot be used.

$$Re_p = \frac{0.1 \times 10^{-4} \times 10}{0.954} = 1.05 \times 10^{-4} < 0.1$$

Stokes Law applies but undoubtedly a Cunningham correction is needed.

$$Kn = \frac{0.0226 \times 2.93 \times 10^{-4}}{0.1 \times 10^{-4} \times 0.5}\sqrt{\frac{573}{28.9}} = 5.89 \quad C = 10.36$$

$$F_D = 3 \times \frac{0.1 \times 10^{-4}}{10.36} \times 2.93 \times 10^{-4} = 2.66 \times 10^{-8} \text{ dynes}$$

c. Deceleration Due to Drag. When no force other than drag is operating, it is obvious that a particle cannot maintain a steady motion relative to the air and that a deceleration must occur. For a spherical particle moving linearly in still air, Newton's law in one-dimension becomes

$$\frac{\pi d_p^{\,3}}{6} \rho_p \frac{du}{dt} = -F_D = -C_D \frac{\pi d_p^{\,2}}{4} \times \frac{\rho_f u^2}{2} \tag{7}$$

This indicates that the deceleration due to drag is

$$\frac{du}{dt} = -\frac{3}{4} C_D \frac{u^2}{d_p} \times \frac{\rho_f}{\rho_p} \tag{8}$$

Before proceeding with the integration of (8), consideration must be given to whether the values of C_D obtained for steady motion will also apply to accelerating (or decelerating) motion. According to Fuchs [10], for Reynolds' numbers not exceeding a few hundred it is possible, without making a large error, to assume that the drag resistance of the air does not depend upon the acceleration. The experimental evidence for this will be discussed below. At this point, the effect of acceleration upon C_D (if any) will be neglected in order that Equation (8) may be solved.

(1) Time of travel. The time of travel required for the velocity to decrease from u_o at $t = 0$, to u may then be found:

$$t = -\frac{4}{3} \frac{\rho_p}{\rho_f} d_p \int_{u_o}^{u} \frac{du}{C_D u^2} \tag{9}$$

To evaluate the integral it must be transformed in order to utilize the relationship between C_D and Re_p. Replacing u by $Re\, \mu / d_p \rho_f$, (9) becomes:

$$t = -\frac{4}{3} \frac{\rho_p d_p^{\,2}}{\mu} \int_{Re_o}^{Re} \frac{d(Re)}{C_D Re^2} \tag{10}$$

(2) Distance of travel. The linear distance x traversed by a particle while the velocity is decreasing from u_o to u may also be found by placing u = dx/dt, replacing dt/u in Equation (8), and integrating. Thus

$$x = -\frac{4}{3}\frac{\rho_p}{\rho_f} d_p \int_{u_o}^{u} \frac{du}{C_D u} \tag{11}$$

Again, transforming the integral into terms of Re,

$$x = -\frac{4}{3}\frac{\rho_p}{\rho_f} d_p \int_{Re_o}^{Re} \frac{d(Re)}{C_D Re} \tag{12}$$

The relationship between distance and time of travel is then

$$\frac{x}{t} = \frac{\mu}{\rho_f d_p} \frac{\displaystyle\int_{Re_o}^{Re} \frac{d(Re)}{C_D Re}}{\displaystyle\int_{Re_o}^{Re} \frac{d(Re)}{C_D Re^2}} \tag{13}$$

Evaluation of the integrals in (10), (12) and (13) depends upon the relationship between C_D and Re. If the effect of acceleration upon drag may be neglected, then the integrals may be calculated using the C_D-Re relationship for steady motion. For this purpose, it is useful to have plots of $1/C_D Re^2$, and $1/C_D Re$, vs. Re to evaluate the integrals. These are shown in Figures 5 and 6.

(3) Drag coefficient-accelerating motion. Torobin and Gauvin [11] have presented an extensive review of the investigations reported on the effect of acceleration upon drag coefficient. They define a modified drag coefficient for accelerating particles as

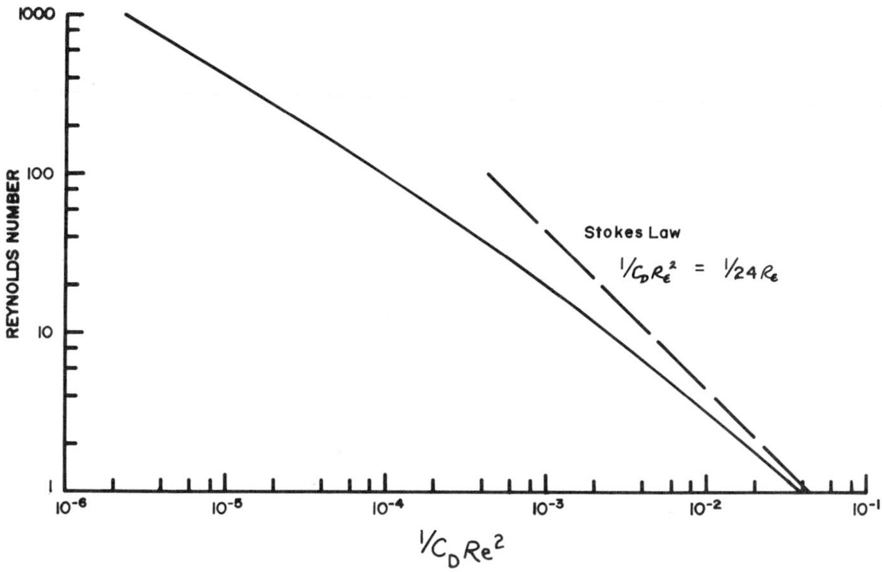

FIG. 5. $1/C_D Re^2$ for Equations (10) and (13).

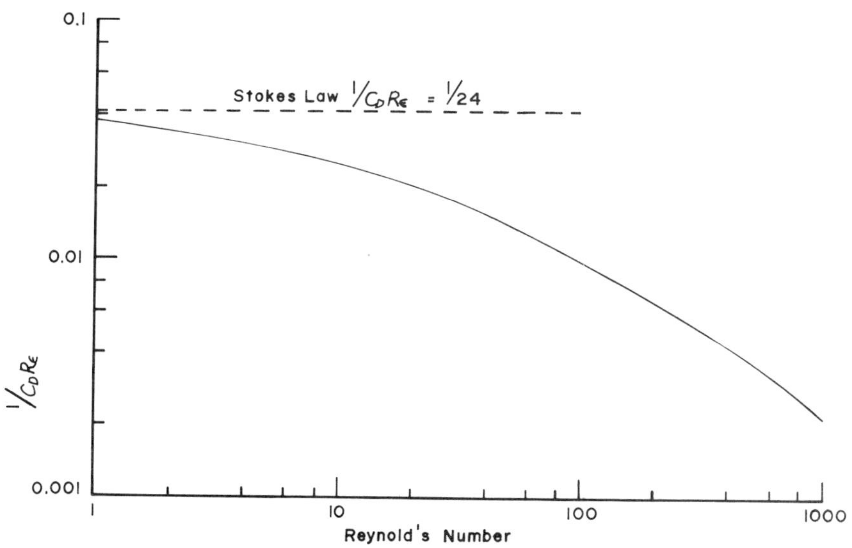

FIG. 6. $1/C_D Re$ for Equations (12) and (13).

$$C_{DA} = \frac{F_{DA}}{\frac{1}{2}\rho_f u^2 \times A_{proj}}$$

where F_{DA} is the drag force acting upon the accelerating particle. At low values of Re_p, experiments indicate that $C_{DA} = C_D$, but at higher values of Re_p some experiments indicate that $C_{DA} > C_D$, others that $C_{DA} < C_D$. There is some indication that $C_{DA} > C_D$ for accelerating particles, and that $C_{DA} < C_D$ for decelerating particles.

For example, the work of Ingebo [12] gives, for deceleration:

$$C_{DA} = 27\ Re_p^{-0.84} \qquad 6 < Re_p < 500$$

in which $C_{DA} < C_D$ throughout the range. However, the experimental evidence is not conclusive, and other effects such as the roughness of the particle and the rate of the acceleration also seem to be involved. Until definitive work has been done on this subject, it is probably satisfactory to follow Fuchs' recommendation, except perhaps in the higher range of Re values encountered in scrubber systems. This question will be considered further in Chapter 9.

(4) Relaxation time and stopping distance. Two very useful concepts emerge from consideration of the deceleration of a Stokes law particle, Equation (8) becomes

$$\frac{du}{dt} = -\frac{18\mu}{\rho_p d_p^2} u = -\frac{u}{\tau} \tag{14}$$

The quantity $\tau = \rho_p d_p^2/18\mu$ is a basic property of the particle-air system which is called the particle relaxation time. From Equations (10) and (12),

$$t = \tau \ln u_o/u \tag{15a}$$

$$u = u_o e^{-t/\tau} \tag{15b}$$

$$x = \tau(u_o - u) - u_o\tau(1 - e^{-t/\tau}) \tag{15c}$$

and for (13)

$$x/t = \frac{(u_o - u)}{\ln(u_o/u)} = u_{\text{log mean}} \tag{15d}$$

The physical significance of τ may be expressed as the time required for the velocity to be reduced by drag to (1/e), or about 36.8%, of its initial value.

The time required for the particle to be brought to rest is infinite, but the distance it travels until u = 0 is finite. It is called the stopping distance of the particle and is given by

$$x_s = \rho_p d_p^2 u_o/18\mu\, u_o = \tau u_o \tag{16}$$

The distance traveled may be expressed as a fraction of the stopping distance, from (15d) in dimemsionless form:

$$x/x_s = (1 - e^{-t/\tau}) \tag{17}$$

These concepts may be extended beyond the Stokes law range. The relaxation time may be defined more broadly from Equation (10) by setting Re = Re_o/e in the integration. The stopping distance may be calculated more generally by setting Re = 0 in the integral of Equation (12). Fuchs [10] has shown how a general chart may be prepared by graphical integration of (12) using the data given in Fig. 6, resulting in

$$\frac{3}{4}\frac{\rho_f}{\rho_p}\frac{x_s}{d_p} = \int_0^{Re_o} \frac{d(Re)}{C_D Re} = f\,(Re_o)$$

The quantity $3\rho_f x_s/4\rho_p d_p$ is plotted as a function of $Re_o = u_o d_p/\nu$. such a chart is given in Fig. 7. In the Stokes Law range $f(Re_o) = Re_0/24$

Example 2. A spherical particle of lead compounds, 100 microns in diameter, density = 6.0 gm/cm^3, is projected horizontally from the tail-pipe of an automobile into still air at 20°C with an initial velocity of 100 cm/sec. How far will it have to travel horizontally, and for how long, when its velocity is 10 cm/sec? What is its stopping distance?

It will be useful to answer the last question first. The stopping distance may be found most readily by use of Fig. 7.

$$Re_o = \frac{d_p u_o \rho_f}{\mu} = \frac{100 \times 10^{-4} \times 100 \times 1.205 \times 10^{-3}}{1.816 \times 10^{-4}} = 6.64$$

This is well out of the Stokes Law range. Hence, read from Fig. 7: $f(Re_o) = 3\rho_f x_s/4\rho_p d_p = 0.24$ corresponding to $Re_o = 6.64$

$$x_s = 0.24 \times \frac{4}{3} \times \frac{6. \times 100 \times 10^{-4}}{1.205 \times 10^{-3}} = 16.0 \text{ cm (stopping distance)}$$

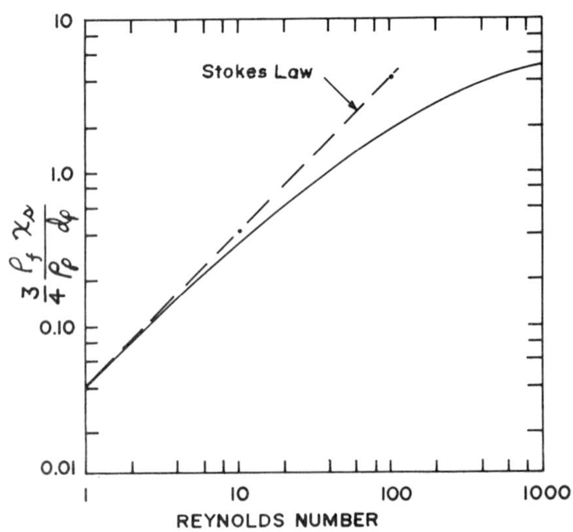

FIG. 7. Stopping distance of particles.

When the velocity of the particle has reached 10 cm/sec, the value of Re will be 0.664. Obtain a value of $f(Re_o)$ corresponding to an Re_o = 0.664 from Fig. 7 and calculate an x_s for it. Subtract this from the value of x_s found above. Thus $3\rho_f x_s / 4\rho_p d_p = 0.027$ when $Re_o = 0.664$ (Fig. 7 extended into Stokes range) and $x_s = 1.80$ cm. Distance traveled = 16. - 1.8 = 14.2 cm.

Finally, the time of flight will be calculated using (10) by performing a numerical integration based upon the data shown in Fig. 5. A plot of $1/C_D Re^2$ vs. Re is constructed and the area under it between Re = 6.64 and Re = 0.664 is obtained.

$$\int_{6.64}^{0.664} \frac{d(Re)}{C_D Re^2} = -0.0769$$

$$t = -\frac{4}{3} \times \frac{6.0(100 \times 10^{-4})^2}{1.816 \times 10^{-4}} (-0.0769) = 0.34 \text{ sec.}$$

Note that in this example only the horizontal motion is considered. At the same time the particle will also be falling under the influence of gravity, which is considered next. However, gravity will have no effect upon the horizontal components of motion.

2. Gravity Force

a. Galileo Number. Because a particle has weight there will always be some tendency for it to move vertically downward. It will also be subjected to a buoyancy force equal to the weight of air it displaces. Although for small particles weight and/or buoyancy may be quite negligible, it is important to consider as a fundamental case of particle motion that in which the forces involved are drag, gravity, and buoyancy.

Newton's Law is written

$$m_p \frac{d\vec{u}}{dt} = \vec{F}_g + \vec{F}_b + \vec{F}_D \tag{18}$$

where $\vec{F}_g = g\ m_p$, the weight of the particle: "gravity force" $\vec{F}_b = g\ m_f$, the weight of displaced fluid: "buoyancy force." Since F_g and F_b act only in the vertical direction, the component equations become

$$m_p \frac{du_x}{dt} = - F_{D_x} \quad \text{horizontal, along x-axis} \tag{18a}$$

$$m_p \frac{du_y}{dt} = g(m_p - m_f) - F_{D_y} \quad \text{vertical, along y-axis taking downward to be the positive direction} \tag{18b}$$

The equation for the horizontal component of motion is the same as that dealt with in the previous section on drag. The second equation, for the vertical component of motion, must now be solved. For spherical particles only:

$$m_p = \pi d_p^3\ \rho_p/6 \quad m_f = \pi d_p^3 \rho_f/6$$

and (18b) becomes

$$\frac{du_y}{dt} = g\left(\frac{\rho_p - \rho_f}{\rho_p}\right) - \frac{F_{D_y}}{m_p} = g\left(\frac{\rho_p - \rho_f}{\rho_p}\right) - \frac{3}{4} C_D \frac{\rho_f}{\rho_p} \frac{u_y^2}{d_p} \tag{19}$$

To place (19) in a dimensionless form, and at the same time to introduce the C_D-Re relationship, replace u_y by $\mu \mathrm{Re}/\rho_f d_p$, and rearrange giving

$$\frac{4}{3} \frac{\rho_p d_p^2}{\mu} \frac{d\mathrm{Re}}{dt} = \frac{4}{3}\ g\ \frac{(\rho_p - \rho_f)\rho_f d_p^3}{\mu^2} - C_D \mathrm{Re}^2 \tag{20}$$

The first term on the right has been called the Galileo Number. It is a property of the fluid-particle system, independent of velocity. It may be defined as:

$$\mathrm{Ga} = \frac{4}{3}\ g\ \frac{(\rho_p - \rho_f)\rho_f d_p^3}{\mu^2} \approx \frac{4}{3}\ g\ \frac{\rho_p \rho_f}{\mu^2}\ d_p^3 \quad \text{for } \rho_f << \rho_p \tag{21}$$

The differential equation for vertical motion may finally be written:

$$\frac{dRe}{dt} = \frac{Ga - C_D Re^2}{24\tau} \tag{22}$$

Taking $u_y = u_{y_o}$, and $Re = Re_o$, at $t = 0$, the solution will involve this integration:

$$\frac{t}{24\tau} = \int_{Re_o}^{Re} \frac{dRe}{Ga - C_D Re^2} \tag{23}$$

A companion equation for the distance travelled vertically downward, may be obtained by placing $dt = dy/u_y$ in (19) and again using $u_y = \mu Re/\rho_f d_p$. Then (19) becomes

$$\frac{dRe}{dy} = \frac{(Ga - C_D Re^2)}{Re} \times \frac{3\rho_f}{4\rho_p d_p} \tag{24}$$

The solution of (24) may be indicated, using the limits $y = 0$ when $u_y = u_{y_o}$, and Re_o.

$$y = \frac{4}{3} \frac{\rho_p}{\rho_f} d_p \int_{Re_o}^{Re} \frac{Red(Re)}{Ga - C_D Re^2} \tag{25}$$

Time of travel and vertical distance of travel may thus be related through (23) and (25).

Both of the integrals involved in these equations must be evaluated by substituting the proper C_D-Re relationship, according to the range of Re encompassed by the limits of integration. Except in the case of Stokes Law this is sufficiently complex that the integrals are best computed by numerical methods. However, as will be shown below, the need to have these integrals evaluated rarely arises in most particle-gas system of interest in dust collection problems. For the sake of completeness, however, the results are presented for Stokes Law particles, where $C_D Re^2 = 24$ Re

$$\frac{t}{\tau} = -24 \int_{Re}^{Re_o} \frac{dRe}{Ga - 24\ Re} = \ln \frac{Ga - 24\ Re_o}{Ga - 24\ Re} \tag{26}$$

$$y = \frac{4}{3} \frac{\rho_p}{\rho_f} d_p \int_{Re_o}^{Re} \frac{Re\ d(Re)}{Ga - 24\ Re} = \frac{\rho_p d_p}{18 \rho_f} \left[(Re_o - Re) - \frac{Ga}{24} \ln \frac{Ga - 24\ Re}{Ga - 24\ Re_o} \right] \tag{27}$$

b. Terminal Settling Velocities. An extremely important aspect of the basic differential equation for vertical motion is the condition when $du_y/dt = 0$. This corresponds also to $dRe/dt = 0$, and represents a state of steady motion. Physically this may be understood as a particle reaching a steady vertically downward velocity, which is called u_s. If the initial downward velocity of the particle $u_{y_o} > u_s$, the particle will decelerate, and if $u_{y_o} < u_s$ it will accelerate, until u_s is attained. The value of u_s is, therefore, called the terminal settling veloctiy of the particle.

Inspection of Equations (22) and (23) reveals that the condition for terminal settling velocity to be attained is that

$$C_D Re^2 = Ga \tag{28}$$

and also reveals that the time required will be infinite. However, as will be shown, for systems of practical interest the limiting condition may be approached very closely within a very short time. For all practical purposes, many such particles may be regarded as settling at their terminal velocities all the time.

To relate the terminal settling velocity to the particle size, the following scheme is used. It is seen that

$$\sqrt[3]{Ga} = \sqrt[3]{C_D Re^2} = \left[\frac{4}{3}g\,\frac{(\rho_p - \rho_f)\rho_f}{\mu^2}\right]^{1/3} d_p \tag{29}$$

is a dimensionless number which is directly proportional to the particle size and does not involve the terminal settling velocity. Similarly,

$$\frac{C_D Re^2}{Re^3} = \frac{Ga}{Re^3} = \frac{C_D}{Re}$$

also satisfies the condition for zero vertical acceleration and produces a dimensionless number

$$\sqrt[3]{\frac{Re_s}{C_D}} = \frac{u_s d_p \rho_f}{\mu} \times \left[\frac{3\mu^2}{4g(\rho_p - \rho_f)\rho_f}\right]^{1/3} d_p = \left[\frac{3\rho_f^{\,2}}{4g(\rho_p - \rho_f)\mu}\right]^{1/3} u_s \tag{30}$$

which is directly proportional to the terminal settling velocity and does not involve the particle size.

From the fundamental data of Fig. 1 or 2 a plot of

$$(C_D Re^2)^{1/3} \text{ vs. } (Re/C_D)^{1/3}$$

may be prepared and is given as Figure 8. Here the ordinate is directly proportional to the value of d_p. For a given fluid-solid system at a specified temperature, both proportionality factors may be calculated and the graph used: (a) to predict the terminal settling velocity of a particle of given size; (b) to determine the size of a particle having a given, or measured, terminal settling velocity. The latter may be used as the basis of an experimental

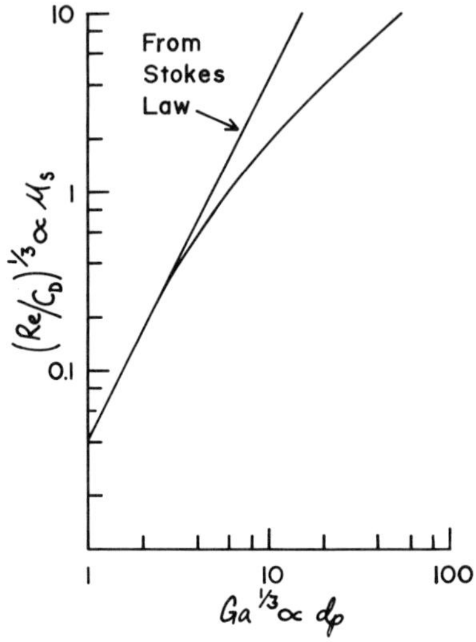

FIG. 8. Terminal settling velocity as related to particle size.

method of determining the "sedimentation diameter" or equivalent spherical diameter, of a particle of unknown or equivalent, shape by measuring its terminal settling velocity.

If greater precision is desired, then the following equations may be used [13].

$$\log\,(Ga^{1/3}) = 0.7841 + 0.7107\,\log\,\left(\frac{Re}{C_D}\right)^{1/3} + 0.1898\left[\log\,\left(\frac{Re}{C_D}\right)^{1/3}\right]^2 + 0.06289\left[\log\left(\frac{Re}{C_D}\right)^{1/3}\right]^3$$

$$\log\,\left(\frac{Re}{C_D}\right)^{1/3} = -\,1.387 + 2.153\,\log\,(Ga^{1/3}) - 0.548\,(\log\,Ga^{1/3})^2 + 0.05665(\log\,Ga^{1/3})^3$$

For a Stokes Law particle the results are simply

$$u_s = \frac{\nu Ga}{24 d_p} = \frac{gC}{18} \frac{(\rho_p - \rho_f)}{\mu} d_p^2 = gC\tau \text{ when } \rho_f << \rho_p \tag{31}$$

or

$$d_p = \sqrt{\frac{18 \mu\, u_s}{gC(\rho_p - \rho_f)}}$$

where the Cunningham factor is placed as shown, when needed. This equation for u_s is sometimes referred to as Stokes Law also.

It must be emphasized that the range of particle sizes for which (31) is accurate, is not the same as that for which Fig. 3 applies. The use of Stokes Law for drag calculation, as given in Fig. 3, is independent of particle density, but the use of (31) to calculate a terminal settling velocity is not independent of ρ_p. The criterion to be applied here is that $Re_s = d_p u_s/\nu \leq 0.1$ or $Ga \leq 2.4$ for accuracy within 2%. Taking $\rho_f << \rho_p$, this corresponds to

$$\frac{4}{3} \frac{g\rho_p \rho_f d_p}{24\mu^2} = \frac{g\tau d_p}{\nu} \leq 0.1$$

In dry air at 20°C and 1 atm., for example, this reduces to

$$\frac{4 \times 980 \times 1.205 \times 10^{-3}}{3 \times 24\ (1.81 \times 10^{-4})^2} \rho_p d_p^{\ 3} = 2.002 \times 10^6 \times \rho_p d_p^{\ 3} \leq 0.1$$

or $\rho d_p^{\ 3} < 49.9 \times 10^{-9}$gm. Table 1 lists values determined from this relationship, and also values for air at higher temperatures.

Example 3. What is the terminal settling velocity of a spherical particle under each of these conditions;

(a) $d_p = 100\mu m$, $\rho_p = 2.60$ gm/cm^3, dry air at 100°C, 1 atm.

(b) $d_p = 60\mu m$, $\rho_p = 1.0$ gm/cm^3, dry air at 300°C, 2 atm.

(c) $d_p = 8\mu m$, $\rho_p = 0.6$ gm/cm^3, dry air at 100°C, 1 atm.

Table 1

Maximum size of sphere for which terminal settling velocity may be calculated by Stokes Law(31)

$Re_s \leq 0.1$ or $Ga \leq 2.4$ for accuracy within 2%
Dry air, at 1 atmos. pressure

	Air Temp. 20°C	Air Temp. 100°C	Air Temp. 300°C
ρ_f =	1.205×10^{-3} gm/cm^3	0.940×10^{-3} gm/cm^3	0.613×10^{-3} gm/cm^3
μ =	1.816×10^{-4} poises	2.17×10^{-4} poises	2.93×10^{-4} poises
$\rho_p d_p^{\ 3}$ =	49.9×10^{-9} gm	92.0×10^{-9} gm	$257. \times 10^{-9}$ gm
ρ_p(gm/cm^3)	d_p(microns)	d_p(microns)	d_p(microns)
0.2	63	77	109
0.3	55	67	95
0.4	50	61	86
0.6	44	54	75
0.8	40	49	68
1.0	37	45	64
1.5	32	39	56
2.0	29	36	50
3.0	26	31	44
4.0	23	28	40
6.0	20	25	35
*Minimum size	8	11	18

*Note: For diameters less than the minimum size listed (taken from Figure 3), a Cunningham correction factor will be needed.

In each case the value of $(Ga)^{1/3}$ should be calculated and used on Fig. 8 to obtain the value of $(Re/C_D)^{1/3}$ from which u_s may be calculated. Table 1 may be used as a quick check to see whether Stokes Law is followed. If so Equation (30) may be used instead of Fig. 8.

(a) From Table 4.3, Stokes law does not apply.

$$Ga^{1/3} = \left[\frac{4}{3} \times 980 \times \frac{(2.6)0.940 \times 10^{-3}}{(2.17 \times 10^{-4})^2}\right]^{1/3} \times 100 \times 10^{-4} = 4.08$$

$(Re/C_D)^{1/3} = 0.55$ from Fig. 8, and Equation (29) gives

$$u_s = 0.55 \times \frac{4 \times 980\ (2.6) \times 2.17 \times 10^{-4}}{3 \times (0.940 \times 10^{-3})^2}^{1/3} = 51.8 \text{ cm/sec.}$$

(b) From Table 4.3, Stokes law applies, even though P = 2 atm, because the value of ρ_f has a negligible effect on Ga. Hence, from (30), C = 1, and

$$u_s = 980 \times 1 \times \frac{1.0 \times (60 \times 10^{-4})^2}{18 \times 2.93 \times 10^{-4}} = 6.7 \text{ cm/sec.}$$

(c) Table 4.3 indicates Stokes law may be used but a Cunningham correction is needed. Fig. 4 gives f = 7.62, hence C = 8.0/7.6 = 1.05 and

$$u_s = 980 \times \frac{1.05 \times 0.6 \times (8 \times 10^{-4})^2}{18 \times 2.17 \times 10^{-4}} = 0.10 \text{ cm/sec.}$$

Example 4. (a) Assuming a raindrop behaves like a rigid sphere what is the size of a drop which is observed to fall at the rate of 5 ft/sec.? (b) A high-volume sampler takes in 60 ft.^{3} of air per minute vertically upward through an area of 1 sq. ft. What is the largest sized particle of density 2.6 gm/cm^3 which it can collect?

(a) Here u_s = 5 x 30.5 = 152 cm/sec. Take the air to be at 20°C, 1 atm.

$$\left(\frac{Re}{C_D}\right)^{1/3} = \left[\frac{3 \times (1.205 \times 10^{-3})^2}{4 \times 980 \times 1.0 \times 1.816 \times 10^{-4}}\right]^{1/3} 152 = 2.78$$

From Fig. 8, $(Ga)^{1/3}$ = 14.2

$$d_p = \left[\frac{3 \times (1.816 \times 10^{-4})^2}{4 \times 980 \times 1.0 \times 1.205 \times 10^{-3}}\right]^{1/3} 14.2 = 0.039 \text{ cm}$$

(b) $$u_s = \frac{60}{1 \times 60} \times 30.5 = 30.5 \text{ cm/sec.}$$

$$\left(\frac{Re}{C_D}\right)^{1/3} = \left[\frac{3 \times (1.205 \times 10^{-3})^2}{4 \times 980 \times 2.6 \times 1.816 \times 10^{-4}}\right]^{1/3} 30.5 = 0.405$$

From Fig. 8, $(Ga)^{1/3} = 3.4$,

$$d_p = \left[\frac{3 \times (1.816 \times 10^{-4})^2}{4 \times 980 \times 2.6 \times 1.205 \times 10^{-3}}\right]^{1/3} 3.4 = 6.8 \times 10^{-3} \text{ cm}$$

c. Gravitational Motion of Stokes Law Particles. The equations for time and distance of settling may be reformulated to advantage in terms of the terminal settling velocities for Stokes law particles. Differential equation (19) will become

$$\frac{du_y}{dt} = g - \frac{u_y}{\tau} \tag{32}$$

Applying the condition that $Ga = 24\ Re_s$ and noting that $Re/Re_s = u_y/u_s$, or $Re_o/Re_s = u_{y_o}/u_s$, Equation (26) for time becomes:

$$\frac{t}{\tau} = \ln \frac{1 - u_{y_o}/u_s}{1 - u_y/u_s}$$

which may be arranged to solve for u_y/u_s

$$\frac{u_y}{u_s} = 1 - \left(1 - \frac{u_{y_o}}{u_s}\right)e^{-t/\tau} \tag{33}$$

For the case of a particle starting to fall from rest $u_{y_o} = 0$, and

$$\frac{u_y}{u_s} = 1 - e^{-t/\tau} \tag{34}$$

Equation (27) for distance becomes

$$y = u_s\ \tau\left[\frac{u_{y_o}}{u_s} - \frac{u_y}{u_s} - \ln \frac{1 - u_y/u_s}{1 - u_{y_o}/u_s}\right] \tag{35}$$

This may be expressed in terms of time rather than velocity by using (33) to replace u_y/u_s. The result is, in dimensionless form:

$$\frac{y}{u_s\tau} = \frac{u_{yo}}{u_s} + (\frac{t}{\tau} - 1) = (1 - \frac{u_{yo}}{u_s})e^{-t/\tau} \qquad (36)$$

When the particle starts from rest

$$\frac{y}{u_s\tau} = \frac{t}{\tau} - (1 - e^{-t/\tau}) \qquad (37)$$

These equations show clearly that when a particle has been falling for a period of time $t \approx 5\tau$ it will have attained over 99.3% of its terminal velocity starting from rest. For most practical purposes, it may be assumed that the acceleration period of fall is essentially completed at $t = 5\tau$, and thereafter the particle continues to fall at steady terminal settling velocity. The additional distance traveled after that point will be virtually equal to $(t - 5\tau)u_s$.

The relaxation time τ is quite small for small particles in air. For example, at 20°C, $\mu = 1.81 \times 10^{-4}$ poises, and for $d_p = 1$ micron $= 10^{-4}$ cm, and for $\rho_p = 1$ gm/cm^3

$$\tau = \frac{1 \times (10^{-4})^2}{18 \times 1.81 \times 10^{-4}} = 3.07 \times 10^{-6} \text{ sec.}$$

and

$$5\tau = 1.54 \times 10^{-5} \text{ sec.}$$

This period of time is so small that for most purposes the accelerating period may be neglected entirely.

d. Dimensionless Forms for Stokes Law Particles. Dimensionless parameters are used extensively in modelling dust collection devices.

Several useful ones may be obtained by putting equation (19) into other interesting dimensionless forms for Stokes Law particles. This is done by selecting various sets of reference values as follows: (i) Select a set of "reference" values, say v_o = an appropriate fixed or constant, velocity; D = an appropriate fixed linear dimension; $t_o = D/v_o$, a reference time. Then define dimensionless variables as $\tilde{u}_y = u_y/v_o$ dimensionless velocity; $\Theta = t/t_o = tv_o/D$ dimensionless time. Substitute these variables into the differential equation which then becomes:

$$\frac{d\tilde{u}_y}{d\Theta} = \frac{g(\rho_p - \rho_f)}{\rho_p}\frac{D}{v_o^2} - \frac{18\mu D}{\rho_p v_o d_p^2}\tilde{u}_y = G - \tilde{u}_y/\mathrm{Stk} \tag{38}$$

Two dimensionless groups appear automatically, and may be defined as

$$G = \frac{g(\rho_p - \rho_f)}{\rho_p}\frac{D}{v_o^2} \simeq \frac{gD}{v_o^2} \quad \text{called "gravity" parameter, which is a dimensionless acceleration} \tag{39}$$

$$\mathrm{Stk} = \frac{\rho_p v_o d_p^2}{18\mu D} = \frac{x_s}{D} \quad \text{the "inertial" parameter, called Stokes Number} \tag{40}$$

which is the ratio of the particle stopping distance to D. Their product is a dimensionless velocity:

$$G \cdot \mathrm{Stk} = \frac{g(\rho_p - \rho_f)d_p^2}{18\mu v_o} = \frac{u_s}{v_o} \tag{41}$$

which is the ratio of the particle terminal settling velocity to v_o. These three dimensionless parameters will be useful in discussing deposition of particles from moving air in the next chapter, where specific meanings will be assigned to v_o and to D.

(ii) Another approach to a dimensionless form may be made by selecting the reference time to be τ, in conjunction with the reference velocity v_o. In this case, the resulting equation is

$$\frac{du_y}{d\Theta} = \frac{g(\rho_p - \rho_f)}{\rho_p} \frac{\tau}{v_o} - \tilde{u}_y = G \cdot Stk - \tilde{u}_y \tag{42}$$

No new groups are obtained.

(iii) Finally, what is perhaps the neatest dimensionless form of equation (19) may be obtained by selecting τ as reference time, $\Theta = t/\tau$ and u_s as reference velocity, $\tilde{u}_y = u_y/u_s$. When these substitutions are made (19) is transformed into

$$\frac{d\tilde{u}_y}{d\Theta} = 1 - \tilde{u}_y \tag{43}$$

which is readily integrated to give

$$\tilde{u}_y = 1 - (1 - \tilde{u}_{y_o})\, e^{-\Theta}$$

This is precisely the same as equation (33) but it is obtained much more easily in this way. This technique of normalizing is often useful in simplifying the solution of differential equations.

3. Other Forces

a. Effect of Other Forces. If forces other than drag, gravity and buoyancy are acting upon the particle, they may easily be incorporated into Newton's Law:

$$m_p \frac{d\vec{u}}{dt} = \vec{F}_D + \vec{F}_g + \vec{F}_B + \Sigma \vec{F}_{other} \tag{44}$$

The nature of the "other force(s)" must be specifically identified, and a functional relationship to all other pertinent variables must be supplied in order to solve the equation and to find the particle trajectory. The treatment here will be limited to the case where other forces are "steady," i.e., independent of particle velocity or or time.

The calculation of the trajectory where an additional constant force is acting is rather simple and depends mainly upon the direction of action of the additional force. Three general cases will be considered first, and then the specific cases of an electrostatic force and a thermal force will be shown.

(a) One-dimensional, vertical motion, F_o vertical; only one component equation exists:

$$m_p \frac{du_y}{dt} = (F_G - F_B) - F_{D_y} \pm F_{o_y} \quad \begin{array}{l} + \text{ if } F_{o_y} \text{ acts downward} \\ - \text{ if } F_{o_y} \text{ acts upward.} \end{array} \tag{45}$$

Since F_{o_y} is constant it may merely be added to $(F_G - F_B)$, and equations (18b), (19), (20) etc. will be essentially the same as before with an added constant term. For steady motion:

$$(F_G - F_B) \pm F_o = F_{D_y} \tag{46}$$

(b) Two-dimensional, vertical and horizontal, F_o is in the same direction as horizontal motion: two component equations are needed:

$$\text{Horizontal:} \quad m_p \frac{du_x}{dt} = F_{o_x} - F_{D_x} \tag{47}$$

(This equation would also cover the case of one-dimensional horizontal motion in which F_{o_x} and F_{D_x} are the only forces acting.)

$$\text{Vertical:} \quad m_p \frac{du_y}{dt} = (F_G - F_B) - F_{D_y} \tag{49}$$

(c) Three-dimensional motion, two F_o forces acting horizontally at right angles: three component equations are needed:

Horizontal: x-direction
$$m_p \frac{du_x}{dt} = F_{ox} - F_{D_x} \tag{49}$$

Horizontal: z-direction
$$m_p \frac{du_z}{dt} = F_{o_z} - F_{D_z} \tag{50}$$

Vertical:
$$m_p \frac{du_y}{dt} = (F_G - F_B) - F_{D_y} \tag{51}$$

Each of these equations is similar in form to one which has been solved above, with the simple addition of the appropriate constant term representing F_o. The condition for steady motion in any one direction is readily obtained by setting the corresponding acceleration term equal to zero.

b. Electrostatic Force. Where a particle is carrying a constant electrical charge and is placed in an electrostatic field of constant potential gradient, an electrostatic force is developed in the direction of the gradient and is given by

$$F_E = q \cdot E \tag{52}$$

where q = charge on particle, coulombs, E = field strength, potential gradient, volts/cm, F_E = electrostatic force, in volt-coulombs/cm, or joules/cm. Wherever F_o appears in the above sets of equations, it would be replaced by F_E where E must be the gradient in the appropriate direction x, y, or z as indicated. Then a new constant term $F_E/m_p = q \cdot E/m_p$, representing acceleration in direction of E, will appear in the corresponding differential equations. The solutions previously obtained may readily be adapted to include F_E. The following example illustrates how this may be done.

Example 6. A Stokes Law particle is projected into still air with an initial horizontal velocity u_{ox}. It immediately enters an electrostatic field of constant horizontal potential gradient E_z, perpendicular to the direction of original motion, and it acquires a charge of q coulombs. Develop equations for determining the location of the particle at any instant after entry. Show the nature of the particle trajectory. Take u_{ox} along x-axis, gravity along y-axis, and $F_{E_z} = q \cdot E_z$ along z-axis, also $u_{oy} = u_{oz} = 0$.

Motion in x-direction: given by equations (14) and (15)

$$x = u_{ox}\, \tau(1 - e^{-t/\tau})$$

Motion in y-direction: given by equations (22) and (36)

$$y = g\tau^2\, \frac{t}{\tau} - (1 - e^{-t/\tau})$$

Motion in z-direction: given by solving

$$\frac{du_z}{dt} = \frac{q_p E_z}{m_p} - \frac{u_z}{\tau}$$

which is in precisely the same form as (31), with qE_z/m_p replacing g. This will give:

$$z = \frac{6q_p E}{\pi\rho_p d_p^{\,3}}\, \tau^2\, \frac{t}{\tau} - (1 - e^{-t/\tau}) = \frac{q_p E\rho_p d_p}{54\pi\mu^2}\, \frac{t}{\tau} - (1 - e^{-t/\tau}) \tag{53}$$

The trajectory will be an exponential-type of curve in three dimensions until $t \simeq 5\tau$. After that time steady conditions will prevail: there will be no further motion in the x-direction, $x = x_s$, and the trajectory will be essentially a straight line in the y-z plane located at $x = x_s$, with

$$u_z = \frac{z}{t} = \frac{6q_p E\tau}{\pi\rho_p d_p^{\,3}} = \frac{qE}{3\pi d_p \mu} = u_{E_s}, \text{ and } u_y = \frac{y}{t} = g\tau = u_s. \tag{54}$$

As we have seen $t = 5\tau$ will be a very short period of time. The in-initial curved protion of the trajectory will be very short, and the overall path will be essentially a striaght line corresponding to the limiting velocity components u_s and u_{E_s} as given.

Attention should be called to the velocity u_{E_s} obtained in the above example. This is a steady-state electrostatic "drift" velocity, which may be obtained by setting the right-hand side of (47) equal to zero. The electrostatic drift velocity will appear again in Chapter 8.

Example 7. Take conditions as in example 6 to be as follows. Particle: size 10μm, density 2.6 gm/cm^3, charge 100 electrons. Field: 2000 volts/cm, air at 20°C; initial velocity: 100 cm/s. Determine the position of the particle with respect to its point of entry into the field, 10 sec. later. Note: charge on a single electron is 1.601×10^{-19} coulombs, and force in volt-coulombs/cm (same as joules/cm) is equal to 10^{-7} times the force in dynes.

The relaxation time is $\tau = 7.98 \times 10^{-4}$ sec., hence $t/\tau = 10/8 \times 10^{-4} \simeq 10^4$ is very large and the acceleration period may be neglected. No Cunningham correction is needed for 10μm particle. The motion is analyzed by components: horizontal motion, x-direction, drag force only:

$$Re_o = \frac{100 \times 10 \times 10^{-4} \times 1.205 \times 10^{-3}}{1.814 \times 10^{-4}} = 0.664$$

Fig. 8 shows this is in Stokes law range. Equation (15) gives:

$$x = x_s = u_o\tau = 100 \times 7.98 \times 10^{-4} = 0.08 \text{ cm}$$

Vertical motion, y-direction, gravity force:

$$Ga = \frac{4}{3}\,\frac{980 \times 2.6 \times 1.205 \times 10^{-3} \times (10 \times 10^{-4})^3}{(1.814 \times 10^{-4})^2} = 0.124$$

$$(Ga)^{1/3} = 0.499$$

Fig. 6 shows this is in Stokes law range. Equation (30) gives:

$$u_s = gC\tau \qquad C=1$$

$$= 980 \times 7.98 \times 10^{-4} = 0.78 \text{ cm/sec.}$$

$$y = u_s t = 0.78 \times 10 = 7.8 \text{ cm}$$

Horizontal motion, z-direction, electrostatic force:

$$u_z = u_{E_s} = \frac{qE}{3\pi d p \mu} \quad \text{(drift velocity)}$$

$$= \frac{100 \times (1.601 \times 10^{-19}) \times 2000 \times 10^7}{3\pi \times (10 \times 10^{-4}) \times 1.814 \times 10^{-4}} = 0.187 \text{ cm/sec}$$

$$z = 0.187 \times 10 = 1.87 \text{ cm}$$

c. Thermal Force-Thermophoresis. It has long been observed that a particle suspended in a gas having a temperature gradient across it will be subjected to a thermal force F_T which causes it to move away from the higher temperature region and toward the lower temperature region of the gas. The process is called thermophoresis. F_T would be be one of the forces represented by F_o in Equations (45) through (51), according to the direction of the temperature gradient.

In the absence of other forces F_T will be opposed only by F_D. Where $F_T = -F_D$ a steady thermophoretic velocity u_T will be reached. Particles will move at this velocity toward a colder surface bounding a hot gas and be deposited upon it. Thermal deposition is a well known phenomena associated with the soiling of walls in the vicinity of hot pipes, and with the fouling of cold surfaces in gas heat exchangers.

An excellent review of theoretical and experimental studies of thermophoresis and thermal deposition has been given by Gieseke

[14]. The thermophoretic velocity may be correlated in terms of the thermal dimensionless group

$$Th = -\frac{u_T \rho_f T}{\mu_f \frac{dT}{dx}} = -\frac{u_T T}{\nu_f \frac{dT}{dx}} \tag{55}$$

in which dT/dx is the temperature gradient (always < 0) in the direction of motion.

For Stokes law particles the value of the thermal force will be given by

$$F_T = \frac{3\pi \mu_f^2 d_p}{C\rho_f T}\left(- Th \frac{dT}{dx}\right) \tag{56}$$

The number Th is found to be essentially constant for small particles, characterized by Kn>>1, indicating that u_T is independent of particle size (and particle properties), directly proportional to the temperature gradient, and only very weakly dependent upon temperature for a given gas. The value of Th, according to various theories seems to lie between 0.42 and 1.5, with experimental values falling in the neighborhood of 0.5.

In this case the force is attributed simply to gas molecules impinging on the particle from opposite sides with different mean velocities. The resulting momentum exchange depends in part upon what fraction of the collisions are perfectly elastic or specular (a = 0), and what fraction are diffuse (a = 1). Experiments indicate that about 80% are diffuse (a = 0.80), which leads to a value of Th = 0.57

For coarser particles, Kn<<2, the situation is more complex and is treated by the thermal creep theory. There is a creeping flow of gas from the colder to the warmer regions along the surface of the particle which in turn experiences a force in the cold direction. Th becomes a function of Kn, as well as of the ratio of the thermal conductivity of the gas to the thermal conductivity of the particle.

Th decreases with Kn in general, and may fall as low as 0.02 at Kn = 0.01 for low k_g/k_p values. Gieseke presents a graph which summarizes all of these relationships for Th [14].

Some idea of the relative importance of thermophoresis may be obtained by estimating u_T in the case of a particle 0.1 micron in diameter in dry air at 300°C and P = 1 atm. From equation (6c) Kn = 2.94. Taking Th = 0.5 for illustration, (55) gives:

$$u_T = \frac{Th \times \nu_f}{T}\frac{dT}{dx} = \frac{0.5 \times 2.93 \times 10^{-4}}{0.613 \times 10^{-3} \times 573}\frac{dT}{dx} = 4.2 \times 10^{-4}\frac{dT}{dx}\frac{cm}{sec}$$

It is evident that unless the temperature gradient is very high, say of the order of 1000°C/cm or more, the thermal velocity will not be large enough to be significant. At least in most situations of ordinary temperature gradients F_T is not important in comparison with other forces acting.

Therefore to use thermal force as a mechanism for particulate collection on a large scale will require a source of heat. This is a major energy requirement. Estimates indicate that the power required may be an order of magnitude greater than that required for other modes of gas cleaning, unless the gas is already hot. In this case, if the gas is cooled by proper heat transfer surfaces thermal deposition will occur. Such deposition from turbulent flows inside tubes and in annular spaces has been studied by Byers and Calvert [15] and by Singh and Byers [16]. Collection efficiencies up to 30% were observed. Appropriate mathematical models were developed. Thermal force may therefore be regarded as a contributing mechanism to particle collection in some situations. Effective thermal precipitators are available to use for collecting small samples of aerosols from flowing gas streams [17]

Because of the relative unimportance of thermal deposition as a mechanism for industrial particle collection, it will not be discussed in further detail. However one should be aware of large temperature gradients wherever they occur, in order not to overlook the possibility of thermal deposition occuring.

II. MOTION DUE TO DIFFUSION

A. Brownian Motion

Aerosol particles suspended in an isothermal gas are constantly subjected to bombardment by the molecules of the gas due to their random thermal motion. At each collision an exchange of momentum occurs. If the aerosol particles are very small these collisions may cause the particles to move about in a random fashion. This is called Brownian motion.

As a consequence of Brownian motion very small aerosol particles tend to diffuse through a gas from regions of higher particle concentration toward regions of lower concentration. The process is analogous to true diffusion in gases and is governed by a differential equation of the same form. In rectangular coordinates, this is

$$\frac{\partial n}{\partial t} = \mathcal{D}\left(\frac{\partial^2 n}{\partial x^2} + \frac{\partial^2 n}{\partial y^2} \quad \frac{\partial^2 n}{\partial z^2}\right) \tag{57}$$

This equation also has the same form as the well known equation for heat conduction and can be treated by the same mathematical techniques.

Particles in Brownian motion will strike a confining surface of the aerosol and may be deposited there. If so, this creates a region of lower concentration adjacent to the surface, and a concentration gradient toward it thus promoting diffusion toward the surface. Diffusive deposition from either stationary or moving gas may, under some conditions, be a significant mechanism of particle collection. It will be discussed further below and in later chapters.

The quantity $\mathcal{D}$ in equation (57), called the diffusivity of the particle, is a basic physical parameter which depends upon the nature and temperature of the gas, and upon the particle size. The value of $\mathcal{D}$ may be estimated in two theoretical ways.

(a) For particles of a size order the same as or greater than the mean free path of the molecules ($Kn \lesssim 0.5$), according to Einstein [18]

$$\mathcal{D} = \frac{CkT}{3\pi \mu d_p} \text{ cm}^2/\text{sec.} \tag{58}$$

where k = Boltzman's constant = 1.37×10^{-16} ergs/°K· molecule.

(b) For particles much smaller than the mean free path of the gas molecules (Kn>>0.5), but still very large compared to the molecules themselves, according to Langmuir [19]

$$\mathcal{D} = \frac{4kT}{3\pi d_p{}^2 p}\sqrt{\frac{8RT}{\pi M}} \tag{59}$$

Symbols are as defined for equation (6c) page 9, but here consistent units must be used.

Illustrative values of $\mathcal{D}$ are shown below for dry air at 20°C and 1 atm. (1.013×10^6 dynes/cm^2). For (58), the values of C were computed by equation (6).

Particle Diffusivities

		$\mathcal{D}$ - cm^2/sec.	
dp-μm	Kn	Einstein (58)	Langmuir (59)
10	0.0131	2.3×10^{-8}	----------
1	0.131	2.7×10^{-7}	----------
0.1	1.31	6.7×10^{-6}	7.8×10^{-6}
0.01	13.1	5.2×10^{-4}	7.8×10^{-4}
0.001	131	----------	7.8×10^{-2}

For very small particles, d_p less than a few tenths of a micron, the Brownian effect becomes significant. It plays a role in particle collection, particle deposition, and particle coagulation. Brownian deposition in aerodynamic capture will be discussed in Chapter 5, and in subsequent chapters where it arises in connection with specific collectors. Particle deposition and coagulation in still air is discussed below.

1. Diffusive Deposition

Equation (57) has been applied to a number of cases of diffusive deposition of aerosol particles from a stationary gas onto confining surfaces in the absence of significant effects from other forces. Fuchs [10, pg. 193] presents results for the following cases: (a) a plane vertical wall in contact with an infinitely large volume of aerosol; (b) a plane horizontal surface, likewise; (c) two parallel vertical planes of infinite extent; (d) two parallel horizontal planes of infinite extent; (e) aerosol situated inside a spherical vessel; (f) aerosol contained inside an infinitely long cylinder; (g) the surface of a sphere surrounded by an infinitely large volume of aerosol; (h) the surface of an infinitely long cylinder, likewise.

In each case diffusion in one dimension only (normal to the surface) is considered, using only the first term on the right side of (57), or its counterpart in spherical or cylindrical coordinates. It is assumed that the aerosol is initially at a uniform concentration n_o of particles all of the same size, and that all undeposited particles remain at the original size, i.e., there is no agglomeration occuring.

The solution of (57) gives first the concentration of remaining particles as a function of location (distance from surface) and time. This is then used in turn to give either the number of particles deposited in time t, or the average concentration $\bar{n}$ remaining undeposited after time t.

For example, in case (e) above, the average remaining concentration is given by

$$\bar{n} = n_o \frac{6}{\pi^2} \sum_{k=1}^{\infty} \frac{1}{k^2} \exp - \pi^2 k^2 \mathcal{D} t/R^2 \tag{60}$$

where R = radius of sphere and k = summation index 1,2, . . . Typical results are tabulated:

$\bar{n}/n_o$	$\mathcal{D}t/R^2$
0.900	0.00093
0.800	0.0039
0.700	0.0094
0.600	0.0180
0.500	0.0305
0.400	0.0485
0.300	0.0744
0.200	0.1135
0.100	0.1830

Thus, particles of 0.1 micron diameter, $\mathcal{D} = 6.7 \times 10^{-6}$ cm^2/sec. in a sphere of 1 mm radius, will be reduced to half their initial concentration after

$$t = \frac{0.0305 \times (0.1)^2}{6.7 \times 10^{-6}} = 45.5 \text{ sec.}$$

Calculations of this kind are not involved in particle collection work, where Brownian motion (if any) is taking place in a moving fluid. However, they may be of interest in estimating deposition in small vessels such as those making up the lungs, during periods of time when there is little or no flow.

2. Thermal Coagulation of Aerosols

Particles in Brownian motion may collide with one another and in so doing adhere to form a new larger particle. This process, called thermal coagulation, is always present in aerosols and will also reduce the concentration of particles as time goes on. If successive collisions occur to the same particles, eventually much larger ones form and may become subject to the action of other forces. Thus coagulation may aid indirectly in the particle collection process. Coagulation may also be promoted by the presence of other forces, e.g., electrostatic, external to the particles and superimposed upon the Brownian movement. An extensive review is given by Zebel [20].

An elementary theory of thermal coagulation, adapted from Smoluchowski [21, 22], is useful to review here. Assuming all particles to be spherical, initially of the same size, and of concentration n per unit volume at any instant, he estimated the total number of collisions per second to be $4\pi d_p \mathcal{D} n^2$. Assuming that each collision results in adherence and therefore reduces the number of particles by one

$$-\frac{dn}{dt} = 4\pi d_p \mathcal{D} n^2 = K_o n^2 \tag{61}$$

Upon integration, assuming K_o to be independent of time,

$$\frac{n}{n_o} = \frac{1}{1 + K_o n_o t}$$

or (62)

$$\frac{1}{n} - \frac{1}{n_o} = K_o t$$

where $n = n_o$ at $t = o$. When $n/n_o = 0.5$, $t = 1/K_o n_o$ is called the half-life of the aerosol. Theoretical values of K_o, the coagulation rate constant, in air at 20°C may be calculated from (61) using the Einstein value for $\mathcal{D}$ from equation (58). Results are:

Particle size d_p-microns	Coagulation rate constant K_o-$cm^3/sec.$
10	2.9×10^{-10}
1	3.4×10^{-10}
0.1	8.4×10^{-10}
0.01	6.5×10^{-9}

It is evident that K_o could not be strictly independent of time, because the average particle size in the aerosol is becoming larger as collisions occur. Nevertheless the reciprocal relationship shown by (62) is found to hold experimentally, albeit with values of K_o some 30% higher than the theoretical. This is attributed to the counter-acting influence of the greater dispersity of particle sizes as coagulation proceeds.

Extension of the theory to the case of particles of two different sizes present initially, shows that if one size is appreciably larger than the other the coagulation rate constant is much greater than that given above. In such aerosols the smaller particles disappear rapidly and are said to be "eaten up" by the larger ones. This is the case with ambient urban aerosols.

Example 8. For particles 0.1 micron in diameter, initially present in air at 20°C, in concentration $10^5/cm^3$, how frequently do collisions occur initially, what is the half-life, and what will the concentration be after 20 minutes?

For $d_p = 0.1$, $K_o = 8.4 \times 10^{-10}$, the initial rate of collision is $K_o n_o^2 = (8.4 \times 10^{-10})(10^5)^2 = 8.4$ per sec.

The half-life is $t_{1/2} = 1/8.4 \times 10^{-10} \times 10^5 \times 3600 = 3.31$ hr.

After t = 20 min.

$$n = \frac{10^5}{1 + \frac{20}{3.31 \times 60}} = 9.08 \times 10^4 \text{ per cm}^3.$$

B. Diffusiophoresis

If a gas is a mixture of two kinds of molecules A and B such that the there is a difference in concentration between two points, the process of gaseous diffusion will take place according to Fick's Law. The mass rate per unit area flow of A will be in the direction of decreasing concentration of A, and proportional to the concentration gradient of A and to the diffusion coefficient of A-B. Similarly for B, except that the concentration gradient (and flux) will be in the opposite direction.

If the molecules of A are heavier than B, then small particles (Kn>>1) suspended in the gas will be subjected to bombardment by a greater proportion of A on one side than the other. The particles will take on a net motion in the same direction as the diffusion of A. This process is called diffusiophoresis.

For the case where the average molecular velocity of the mixture is zero (A and B counter-diffusing through each other), according to Waldmann and Schmitt [23], the particle takes on a diffusion velocity u_D in the x-direction, given by

$$u_D = \left[\frac{\delta_A\sqrt{m_A} - \delta_B\sqrt{m_B}}{\delta_A n_A\sqrt{m_A} + \delta_B n_B\sqrt{m_B}}\right] D_{AB}\,\frac{\partial n_A}{\partial x} \tag{63}$$

in which m = mass of molecule, proportional to molecular weight; n = concentration of molecule, or mole-fraction; $\delta = 1 + \frac{\pi}{8}a$; a = accomodation coefficient; D_{AB} = mutual diffusion coefficient for the two gasses. The velocity is in the direction of the gradient of $A(\partial n_A/\partial x < 0)$, the heavier molecule. Since $a_A \simeq a_B$, (63) simplifies to

$$u_D = -\left[\frac{\sqrt{m_A} - \sqrt{m_B}}{m + \sqrt{m_A m_B}}\right] D_{AB} \frac{\partial n_A}{\partial x} \tag{64}$$

in which $m = n_A m_A + n_B m_B$, is proportional to the average molecular weight of the mixture. Note that in air where $m_{O_2} \simeq m_{N_2}$, the quanity in brackets is very small (0.0062) and the concentration is uniform so that u_D is ordinarly zero. In a mixture for which $m_A = m_B$, u_D would always be zero, regardless of concentration gradients.

For the special, but important, case where one gas is stationary overall, another phenomenon comes into play. Consider the surface of an evaporating liquid which is sending vapor molecules to diffuse into the surrounding resting gas. To compensate for the tendency of the gas molecules to diffuse toward the surface, there must be a (hydrodynamical) flow of gas-vapor mixture away from the surface at a velocity u_{St} given by

$$n_A u_{St} = D_{AB}\, \partial n_A/\partial x \tag{65}$$

where x is taken as a positive direction away from the surface. This velocity is called a Stefan flow. It will be superimposed upon the diffusion velocity, so that the net velocity is $(u_D + u_{St})$.

Adding (64) and (65), in which A refers to gas molecules and B to vapor, and using the fact that $\partial n_A/\partial x = -\,\partial n_B/\partial x$ it may be shown that

$$u_D + u_{St} = -\frac{\sqrt{m_B}}{n_A\sqrt{m_A} + n_B\sqrt{m_B}}\,\frac{D_{AB}}{n_A}\,\frac{\partial n_B}{\partial x} \tag{66}$$

The net result is always a particle velocity away from the evaporating surface, as $\partial n_B/\partial x$ is < 0. On the other hand, if the vapor is condensing from the gas onto a liquid surface the particle motion will be toward the surface, as $\partial n_B/\partial x$ is > 0.

A diffusion force F_D, analogous to the thermal force F_T given by equation (56), may be calculated for Stokes law particles in the same way.

$$F_D = 3\pi\mu d_p(u_D + u_{St})/C \tag{67}$$

Well-known observations confirm these predictions. A clear space forms in a dusty gas above the surface of an evaporating drop of liquid. Experiments confirm the Stefan-Waldmann theory quantitatively [24].

The phenomenon is of importance in wet scrubbing systems of particulate collection where drops of colder liquid are used as target collectors. If the surrounding gas is saturated with liquid, condensation will occur and thus help to drive dust particles onto the target drops. This is referred to as force-flux-condensation (FFC). Conversely, if the gas is unsaturated and warmer drops are evaporating, the particle collection process is hindered. The role of FFC will be discussed further in Chapter 9.

REFERENCES

1. A.H. Shapiro, *Shape and Flow,* Anchor Books, Doubleday and Co., Inc., New York, (1961).

2. J.H. Perry ed., *Chemical Engineers Handbook,* 5th edition McGraw Hill, New York, (1976).

3. G.G. Stokes, *Trans. Cambridge Phil. Soc.,* 9: Part II, 51 (1851).

4. C. Oseen, *Neure Methoden und Ergebnisse in der Hydrodynamik,* sec. 16, Leipzig, (1927).

5. L. Schiller and L.A. Naumann, *Zeit, Vor. Deut. Ing.,* 77: 318 (1933).

6. L. Klyachko, *Otopl. i Ventil.*: No. 4 (1934).

7. D.R. Dickinson and W.R. Marshall, *A.I.Ch.E. Journ.*, 14: 541 (1968)

8. C.N. Davies, *Proc. Phys. Soc.*, 57: 259 (1945).

9. S. Calvert in *Air Pollution-Vol. IV*, 3rd Ed. (A. Stern, ed.), Academic Press, New York (1977).

10. N.A. Fuchs, *The Mechanics of Aerosols*, Pergamon Press, New York (1964).

11. L.B. Torobin and W.H. Gauvin, *Canad. J. Chem. Eng.*, 37: 224 (1959).

12. R. Ingebo, *NASA:* Tech. Note 3762 (1956).

13. W. Koch, Personal Communication (1977).

14. J.A. Gieseke, in *Air Pollution Control-Part II* (W. Strauss, Ed.) Wiley Interscience, New York (1972).

15. R.L. Byers and S. Calvert, *Ind. Eng. Chem. Fund.*, 8: 646 (1969)

16. B. Singh and R.L. Byers, *Ind. Eng. Chem. Fund.*, 11: 127 (1972).

17. T. Allen, *Particle Size Measurement*, 2nd ed. pg. 62, Chapman and Hall, London, (1975)

18. A. Einstein, *Z. Electrochemie*, 14: 235 (1908).

19. I. Langmuir, *OSRD* Report No. 865 (1942).

20. G. Zebel, in *Aerosol Science* (C.N. Davies, Ed.) Chap. II, Academic Press, New York (1966).

21. M. von Smoluchowski, *Phys. Z.*, 17: 557, 585 (1916).

22. M. von Smoluchowski, *Z. Phys. Chem.* 92: 129 (1918).

23. L. Waldman and K.H. Schmitt in *Aerosol Science* (C.N. Davies, Ed.) Chap. VI, Academic Press, New York (1966).

24. P. Goldsmith and F.G. May in *Aerosol Science* (C.N. Davies, Ed.) Chap. VII, Academic Press, New York (1966).

25. K. Willeke, *J. Aerosol Sci.*, 7: 381 (1976).

CHAPTER 5

TRAJECTORIES OF PARTICLES IN MOVING GAS - AERODYNAMIC CAPTURE

I. PARTICLE TRAJECTORIES

This chapter deals in detail with the methods of predicting whether a particle being carried in a moving stream of gas will collide with a collecting surface or with a target collector. It is based upon the general discussion of Mechanisms of Collection given in Chapter 3. The basic approach is to calculate the trajectory or path followed by a given particle from its point of entrance into a zone of collection under whatever forces may be present to act upon it. The information presented in Chapter 4 about forces and motion of particles in a still gas will be applied here to a flowing gas. The results obtained will in turn form the basis for the modelling of the various kinds of collectors which are treated in the subsequent chapters.

A. The Role of the Gas Velocity

The prediction of particle trajectories in a gas which is in motion is more complex than for a gas which is at rest. The velocity of the gas stream with respect to the particles, as well as with reference to the stationary surroundings, must be taken into account. A differential equation must be written for the streamlines of the gas as well as for the aerosol particle, and these two equations considered simultaneously. Velocity vectors for the gas, and for the particle, must be determined.

Whenever the velocity vector for the fluid at a point differs in direction and/or magnitude from that of the particle at the same location, there is said to be "slip" between particles and gas. The vector difference of these velocities is called the "slip," or "relative," velocity. The magnitude of this relative velocity determines the magnitude of the drag force. The equations of Chapter 4 for calculating drag are applicable so long as relative velocity is substituted for particle velocity.

In Newton's Law the expression for the drag force will become

$$\vec{F}_D = -\frac{C_D A_{proj} \rho_f}{2} (\vec{u} - \vec{v}) \cdot |\vec{u} - \vec{v}| \tag{1}$$

$$|\vec{F}_D| = \frac{C_D A_{proj} \rho_f}{2} |\vec{u} - \vec{v}|^2 \tag{2}$$

where $\vec{u}$ = velocity vector of particle; $\vec{v}$ = velocity vector of fluid. Figure 1 illustrates these vectors and their components, and shows that $(\vec{u} - \vec{v})$ = velocity vector of particle relative to fluid and

$$|\vec{u} - \vec{v}| = \sqrt{(u_x - v_x)^2 + (u_y - v_y)^2} \tag{3}$$

In calculations, this drag force will have to be resolved into its components:

$$F_{D_x} = \vec{F}_D \cos \alpha = \frac{C_D A_{proj} \rho_f}{2} (v_x - u_x) |\vec{u} - \vec{v}| \tag{4}$$

$$F_{d_y} = \vec{F}_D \sin \alpha = \frac{C_D A_{proj} \rho_f}{2} (v_y - u_y) |\vec{u} - \vec{v}| \tag{5}$$

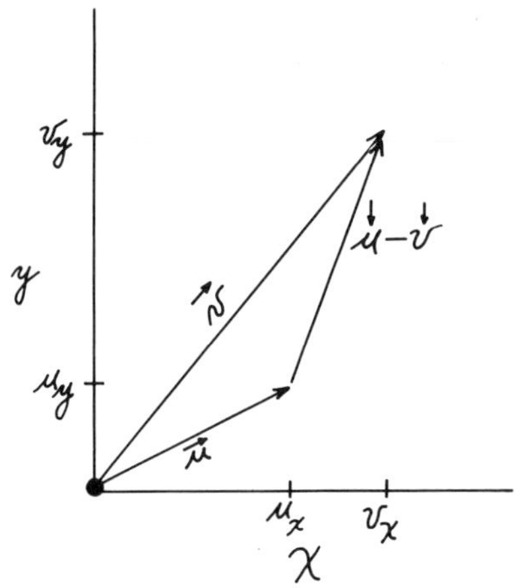

FIG. 1. Fluid velocity $\vec{v}$, particle velocity u, and drag ($\vec{u}$ - $\vec{v}$) vectors.

Several basic cases may be examined, with reference to the concepts of collecting zones, collecting velocity, and residence time described in Chapter 3. The simplest is that of plug flow (constant $\vec{v}$) for the fluid, and a constant collecting velocity $\vec{u}$ perpendicular to $\vec{v}$. As illustrated in Fig. 7, Chapter 3, the particle velocity is a constant and the trajectory is a straight line.

Consider the same case but with laminar flow of the gas stream. The fluid will have a symmetrical parabolic velocity distribution profile across the zone. As the particle moves across the fluid it will constantly encounter a different value of v, hence its velocity vector will vary in length and direction. The path of the particle will no longer be straight. The actual trajectory may be visualized readily, Figure 2, by comparing it with the straight-line path the particle would have followed in a plug-flow air stream at $v = v_o$, where v_o is air velocity at the point of entry of the particle. If it can be assumed that the particle instantaneously adjusts its velocity u_x to be equal to v_x at every point, i.e., no slip in the x-direction, the curved path may be calculated from a knowledge of the actual fluid velocity profile and the nature of the collecting force. This is done below for certain cases.

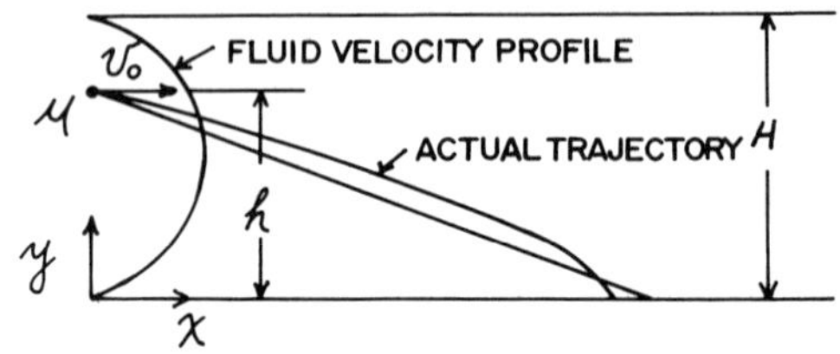

FIG. 2. Trajectory of particle between parallel plates, laminar flow.

The precise shape of the fluid velocity profile will vary with the value of Reynold's number based upon the duct size. When this number becomes sufficiently large, there will be an onset and development of turbulent flow. Not only will the gross velocity profile become flatter and less parabolic, but also there will now be eddies in the fluid to affect the particle motion.

Another case of importance is that of the flow of the gas stream around a collector target element. (See Fig. 1 (b), Chapter 3). Here a more complex pattern of fluid stream lines will be set up around the curvature and into the wake of the target. The exact nature of the pattern will vary with fluid Reynold's number based upon the size of the target, i.e., $Re_f = vD/\nu_f$ Where these streamlines begin to curve upstream of the target, the particle may not be able to follow owing to its inertia. The particle path will be a curve cutting across the fluid streamlines toward the target surface, as shown in Figure 3. There will be "slip," and the drag force must be reckoned accordingly. A knowledge of the fluid mechanics of flow

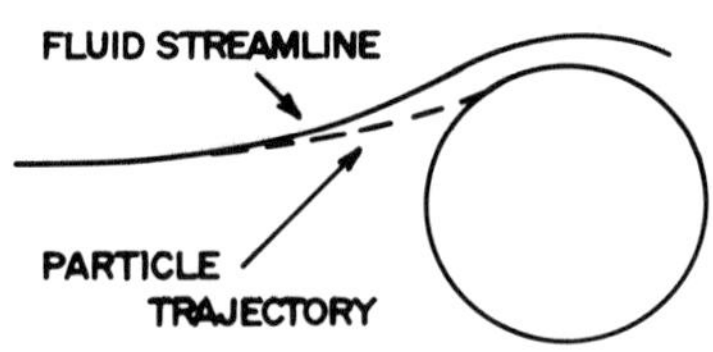

FIG. 3. Trajectory of particle approaching target surface.

around whatever shape the target is will be needed in order to place the fluid streamlines.

These examples illustrate the importance of the fluid velocity in determining the particle trajectory. Particle trajectories will next be examined in detail with the objective of determining deposition upon collector surfaces (parallel planes, and cylinders) and upon collector targets (cylinders and spheres).

B. Trajectories Within Containment Zones

1. Parallel Plane Boundaries

If the collection zone is bounded by parallel plane surfaces as in Fig. 1a, Chapter 3, and the collecting velocity $\vec{u}$ is constant, the particle trajectory will depend upon the velocity profile of the gas flow between the planes. There are three cases of interest: (a) plug flow, $\vec{v}$ constant; (b) laminar flow, $\vec{v}$ parabolic profile; (c) turbulent flow, $\vec{v}$ logarithmic profile, eddy currents. Some examples are: (1) gravity settling between horizontal planes, where $\vec{u}$ acts vertically downward and is equal to u_s, the terminal settling velocity of the particles; equations (29) or (31), Chapter 4. (2) electrostatic attraction toward a plane bounding a potential difference across a gas, with $\vec{u}$ acting in the direction of the potential gradient, and equal to u_E; equation (34), Chapter 4. (3) thermophoresis between two planes of different temperatures, $\vec{u}$ acting from hotter toward colder plane and equal to u_T, the thermophoretic velocity of the particle; equation (55), Chapter 4. (4) diffusiophoresis from the horizontal surface of an evaporating liquid (Stefan flow), $\vec{u}$ acting vertically upward equal to u_D; equation (63) or (65), Chapter 4.

These and similar examples may all be analyzed in one general treatment making these assumptions:

- The aerosol composed of various sized particles (i-grade) is homogeneous and uniformly distributed between planes at the point of entrance into the zone.

- The particles behave individually, acted upon by the collecting force(s) and drag; buoyancy is neglected.
- There is no slip in the direction of flow.
- Particles accelerate rapidly to the collecting velocity, which is constant in the direction perpendicular to flow, and acts only toward one plane.
- The planes may be horizontal or vertical; if horzontal, gravity will have to be considered as part of the total collecting force (action toward the lower plate only).

As shown in Fig. 2, the coordinate system is taken as x in the direction of flow from point of entrance, y perpendicular to flow from plane toward which collecting velocity acts. The spacing between the planes is H, their length in direction of flow is ℓ. The point at which a particle enters the zone is at x = 0, y = h.

a. Plug Flow of Gas.

Fluid: $v_x = v_o$, $v_y = 0$

Particle $u_{x_i} = v_o$ (independent of d_{p_i})

$u_{y_i} = u_{c_i}$, constant but may depend upon d_{p_i}

This corresponds to the no-mixing model presented in Chapter 3, III.B.1, illustrated there with $u_{c_i} = u_{s_i}$ the terminal settling velocity of particles between horizontal planes.

The particle trajectories are straight lines having slopes = u_{c_i}/v_o, and the grade efficiency is given by

$$\eta_i = u_{c_i}\ell/v_oH \qquad (6)$$

All particles for which $u_{c_i} \geqq v_oH/\ell$ will be collected with 100% efficiency. This establishes a minimum particle size which will be collected completely.

The relationship between η_i and d_{p_i} will depend upon the nature of the collecting force and its relationship to d_{p_i}. This can be obtained from Chapter 4. For example:

(a) Gravity settling: $u_{c_i} = u_{s_i}$ which depends upon Ga_i and may be obtained from Fig. 9, Chapter 4. For Stokes law particles $u_{s_i} \propto d_{p_i}{}^2$ and the grade-efficiency plot is as shown in Fig. 8, Chapter 3.

(b) Electrostatic attraction: $u_{c_i} = u_{E_i}$, given by equation (34), Chapter 4. For Stokes law particles $u_{E_i} \propto d_{p_i}$, and the grade-efficiency plot is linear from $\eta_i = 0$ at $d_{p_i} = 0$, up to $\eta_i = 1$ at d_{p_i} corresponding to $u_{E_i} = v_o H/\ell$.

(c) Thermophoresis: $u_{c_i} = u_{T_i}$, given by equation (55), Chapter 4, which is independent of d_{p_i} for Kn >> 1. The grade-efficiency is the same for all particle sizes and is given by (6) with $u_{c_i} = u_T$.

b. Laminar Flow of Gas.

$$\text{Fluid:}\quad v_x = v_{max}\left[1 - \frac{4(y - H/2)^2}{H^2}\right] = 4v_{max}\left[\frac{y}{H} - \left(\frac{y}{H}\right)^2\right]$$

$$v_y = 0$$

$$v_{max} = v_{max} = 3/2v_o \qquad (v_o = \text{average velocity})$$

$$\text{Particle:}\quad u_{x_i} = v_x \quad (\text{independent of } d_{p_i})$$

$$u_{y_i} = u_{c_i}, \text{ constant, but may depend upon } d_{p_i}$$

The parabolic fluid velocity profile, illustrated in Fig. 2, holds for $Re_f = v_o H/\nu_f < 890$.

To obtain the particle trajectory set

$$u_{x_i} = v_x = \frac{dx}{dt} \quad \text{and} \quad u_{y_i} = u_{c_i} = -\frac{dy}{dt}$$

Then

$$\left(\frac{dx}{dy}\right)_i = -\frac{6v_o}{u_{c_i}}\left[\frac{y}{H} - \frac{y^2}{H}\right] \tag{7}$$

It is convenient to put (7) into dimensionless form using H and v_o as reference values. Then $x^* = x/H$; $y^* = y/h$; $u_{c_i}{}^* = u_{c_i}/v_o$ and (7) becomes

$$\left(\frac{dx^*}{dy^*}\right)_i = -\frac{6}{u_{c_i}^*}\left(y^* - y^{*2}\right) \tag{8}$$

Integrating (8) with the boundary condition that at $x^* = 0$, $y^* = h/H$ $y^* = h/H = h^*$, gives the equation of the trajectory defined for $0 \leq y^* \leq h^*$, i.e., from point of particle entrance at y = h, to point where particle collides with collecting plane at y = 0. The result may be written

$$x^* = \frac{1}{u_{c_i}{}^*}\left[\left(2y^{*3} - 3y^{*2}\right) - \left(2h^{*3} - 3h^{*2}\right)\right] \tag{9}$$

All types of trajectories represented by (9) have a point of inflection at $y^* = 1/2$, and are normal to the surface at y = 0.

The efficiency may be determined by finding the value of h^* for the trajectory which terminates at y = 0, x = ℓ or $x^* = \ell/H$. Equation (9) gives

$$3h_i{}^{*2} - h_i{}^{*3} = \frac{\ell}{H}\frac{u_{c_i}}{v_o} \tag{10}$$

All particles entering below this value of $h_i{}^*$ are collected, hence $\eta_i = h_i{}^* =$ root of (10) lying between 0 and 1. The values of this root corresponding to values of $\ell u_{c_i}/Hv_o$ are tabulated below.

u_{c_i}/Hv_o	h_i^*	$\ell u_{c_i}/Hv_o$	h_i^*
0	0	0.6	0.5671
0.1	0.1958	0.7	0.6367
0.2	0.2871	0.8	0.7129
0.3	0.3633	0.9	0.8042
0.4	0.4329	1	1
0.5	0.5		

c. Turbulent Flow of Gas. As the value of $Re_f = v_oH/\nu_f$ for the duct increases above about 890, there will be an onset of turbulence in the fluid. The eddy currents developed will attain root-mean-square velocities in the vertical direction of approximately 0.03 to 0.1 of v_o, the average fluid velocity. For example, at $v_o \simeq 1$ m/sec, these will be 3-10 cm/sec, appreciably greater than the terminal settling velocity of say 20-micron particles of unit density, $u_s \approx 1.3$ cm/sec. Hence particles of this size and finer will tend to become uniformly mixed over the vertical cross-section of the duct at velocities $v_o \geq 1$ m/sec.

When such conditions prevail, the collection efficiency may be modelled by the lateral-mixing model discussed in Chapter 3, III.B.2. The velocity profile becomes very flat and $v \simeq v_o$. In time interval $dt = d\ell/v_o$ the reduction in number of particles of ith grade by collection from an element of gas $d\ell$ in length and unit width, will be $-dn_i = (nu_{c_i}/H)$ at $= (nu_{c_i}/Hv_o)\, d\ell$. The parameter $k_i = u_{s_i}/Hv_o$, and the collection efficiency is given by

$$\eta_i = 1 - \exp - (u_{c_i}\ell/Hv_o) \tag{11}$$

In all of the above equations (6), (10), and (11) for the grade-efficiency of collection by a constant collection velocity, there are two parameters. One is the velocity ratio $u_{c_i}^* = u_{c_i}/v_o$ which is determined by the properties of particle and fluid, and by operating conditions. The other is geometric ratio ℓ/H which represents the physical shape of the collector space. It will be seen that these two kinds of parameters will always occur in the model for any collecting device.

Comparison of grade-efficiencies from (6), (10) and (11) shows that for $0 < u_{c_i}\ell/v_o H < 0.5$ they rank in the order LAMINAR > PLUG > TURBULENT, while for $0.5 < u_{c_i}\ell/v_o H$ the order is PLUG > LAMINAR > TURBULENT. At $u_{c_i}\ell/v_o H \geq 1$, $\eta_i = 1$ for PLUG and LAMINAR, but $\eta_i \to 1$ asymptotically ($\eta_i = 0.9933$ at $u_{c_i}\ell/v_o H = 5$) for TURBULENT.

Example 1. A thermal sampling collector, consisting of parallel horizontal plates, operates under the following conditions: $H = 2$mm, $\ell = 5$ cm; top plate at 100°C, bottom plate at 0°C; $v_o = 10.0$ cm/s; dry air at 1 atm. Estimate the grade efficiency of collection of particles of unit density ranging from $d_p = 0.001$ µm to 1 µm.

Assuming the average gas temperature to be 50°C, $\mu = 1.96 \times 10^{-4}$ gm/cm·s, $\rho_f = 1.093 \times 10^{-3}$ gm/cm^3, and $v_o H/\nu_f = 11.1 < 890$, so that flow is laminar.

The value of Kn (equation 6c, Chapter 4) will range from 148 to 0.148, so that Th may be taken as ≈ 0.50 and u_T will be

$$u_T = \frac{0.50 \times 1.96 \times 10^{-4}}{1.093 \times 10^{-3}\ (273 + 50)} \times \frac{(100 - 0)}{(0.2)} = 0.14 \text{ cm/s}$$

In equation (10), u_T will be the same for all particles

$$\frac{\ell}{H}\frac{u_T}{v_o} = \frac{5 \times 0.14}{0.2 \times 10.0} = 0.35$$

and $h^* \simeq 0.40$. Particles of all sizes in this range will be collected by a thermal force with an efficiency of 40%. If it is desired to obtain 100% collection efficiency, the air flow rate should be reduced to below

$$v_o = \frac{5 \times 0.14}{0.2 \times 1} = 3.50 \text{ cm/s}$$

There will also be a small effect of gravity, so that the coarser particles will tend to be collected to a slightly greater extent. The terminal settling velocity of 1 μm particles is given by equation (31), Chapter 4, as

$$u_s = \frac{980 \times 1.18 \times (1)}{18 \times 1.96 \times 10^{-4}} \left(1 \times 10^{-4}\right)^2 = 0.0033 \text{ cm/s}.$$

Since $u_s << u_T$ the gravity effect may be neglected.

2. Cylindrical Boundary-Spinning Gas

Centrifugal force may be utilized as an effective collecting force by causing the dust-laden gas stream to flow in a circular manner inside a cylindrical boundary. This is the basis for the operation of cyclone collectors which are discussed in detail in Chapter 6. Fundamental to the modelling is a knowledge of particle trajectories in a spinning gas which is presented here.

Consider a particle entering tangentially at speed u_{T_1} into a horizontal plane of a spinning gas stream at R_1 as shown in Figure 4. Owing to its inertia, which is represented by a centrifugal force of $F_c = m_p u_T^2/R$ at any instant and to the force of gravity, the particle will follow a helical path downward. Its velocity vector u will have a tangential component (u_T), a radial component (u_R), and an axial component (u_s). The velocity of the spinning gas will be assumed to

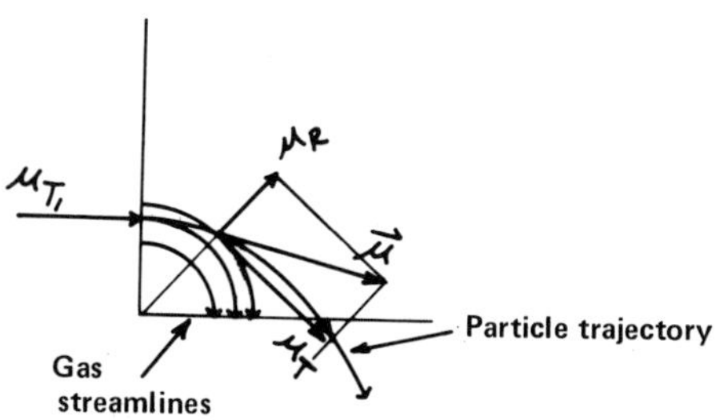

FIG. 4. Particle trajectory in a spinning gas.

have only a tangential component v_T, with $v_R = 0$. It may also be assumed that there is no tangential slip, i.e., that $u_T = v_T$. However, v_T is not to be regarded as a constant, but to vary with R. Tests have shown that this variation will likely follow the relationship

$$u_T R^n = \text{constant} = u_{T_1} R_1^{\,n} \tag{12}$$

For an ideal fluid in such a vortex flow n = 1, but in real flows the value of n may range downward to 0.5.

Newton's Law is applied only to the radial component of motion, with the drag force directed radically inward.

$$m_p \frac{du_R}{dt} = \frac{m_p u_T^{\,2}}{R} - F_D \tag{13}$$

Since $u_R = \frac{dR}{dt}$, this may be written for a spherical particle:

$$\frac{d^2R}{dt^2} = \frac{u_{T_1}^{\,2} R^{2n}}{R^{2n+1}} - \frac{3}{4} \frac{C_D}{d_p} \frac{\rho_f}{\rho_p} \left(\frac{dR}{dt}\right)^2 \tag{14}$$

This equation is obviously non-linear and is further complicated by the fact that C_D will also vary with u_R.

Applying (14) to a Stokes law particle gives some simplification:

$$\frac{d^2R}{dt^2} = \frac{u_{T_1}^{\,2} R_1^{\,2n}}{R^{2n+1}} - \frac{1}{\tau} \frac{dR}{dt} \tag{15}$$

but (15) cannot be solved in general except by numerical approximation methods. Fortunately a good approximation to it is obtained by neglecting the second order term although this is equivalent to saying that the particle moves radially outward at a constant velocity which is obviously inconsistent with the resulting equation

$$\frac{dR}{dt} = \frac{\tau u_{T_1}{}^2 R_1{}^{2n}}{R^{2n+1}} = \frac{\tau u_{T_2}{}^2}{R_2} \left(\frac{R_2}{R}\right)^{2n+1} \tag{16}$$

This may be integrated to give the position R at any time t with respect to R_1 when t = 0.

$$t = \frac{R_1{}^2}{2(n+1)\tau u_{T_1}{}^2} \left[\left(\frac{R}{R_1}\right)^{2n+2} - 1 \right] \tag{17}$$

This equation relating radial position R to elapsed time t enables calculation of the particle trajectory. It will be used as the basis for modelling the cyclone collector in Chapter 6. It is useful to restate it in terms of u_{T_2} and R_2, using (11),

$$t = \frac{R_2{}^2}{2(n+1)\tau u_{T_2}{}^2} \left[\left(\frac{R}{R}\right)^{2n+2} - \left(\frac{R_1}{R_2}\right)^{2n+2} \right] \tag{18}$$

where R_1 and R_2 designate respectively the innermost and the outermost radial positions of interest, and $R_1 < R < R_2$.

C. Inertial Trajectories

The path of a particle moving through a spinning gas as just discussed, is a special case of a trajectory determined by particle inertia as the gas flow is subjected to a curved path. A more general approach will be considered next, which may be used to treat several other cases of importance in the modelling of dust collectors, especially for target collectors.

For this purpose, only the drag force acting upon the particle is considered and Newton's law is written using (1)

$$m_p \frac{d\vec{u}}{dt} = \vec{F}_D = \frac{-C_D A_{proj} \rho_f}{2} (\vec{u} - \vec{v}) \cdot |\vec{u} - \vec{v}| \tag{19}$$

and placed in dimensionless form by using set (a) of the reference values discussed in Chapter 4, equation (38). Here v_o will represent an undisturbed steady velocity at some point upstream of onset of the curved path (upstream of target), D will represent some appropriate dimension related to the size of the curvature in path, (size of target), and the reference time will be D/v_o. The dimensionless variables then become

$\theta = \frac{v_o}{D} t$ dimensionless time

$\tilde{x} = \frac{x}{D}$ dimensionless particle position along x-axis

$\tilde{y} = \frac{y}{D}$ dimensionless particle position along y-axis

$\tilde{u}_x = \frac{u_x}{v_o} = \frac{1}{v_o} \frac{dx}{dt}$ dimensionless particle velocity, x-component

$\tilde{u}_y = \frac{u_y}{v_o} = \frac{1}{v_o} \frac{dy}{dt}$ dimensionless particle velocity, y-component

$\tilde{v}_x = \frac{v_x}{v_o}$ dimensionless fluid velocity, x-component

$\tilde{v}_y = \frac{v_y}{v_o}$ dimensionless fluid velocity, y-component

Since there is usually symmetry about the z-axis, it is sufficient to deal with the component equations in two dimensions (x, y)

only. Equation (19) then is resolved into the scalar equations using (4) and (5):

$$m_p \frac{du_x}{dt} = - C_D A_{proj} \frac{\rho_f (u_x - v_x)(|\vec{u} - \vec{v}|)}{2}$$
$$m_p \frac{du_y}{dt} = - C_D A_{proj} \frac{\rho_f (u_y - v_y)(|\vec{u} - \vec{v}|)}{2} \qquad (20)$$

which are reexpressed in dimensionless form:

$$\frac{d\tilde{u}_x}{d\theta} = - \frac{C_D A_{proj} D \rho_f}{2m_p} (\tilde{u}_x - \tilde{v}_x) \frac{(|\vec{u} - \vec{v}|)}{v_o}$$
$$\frac{d\tilde{u}_y}{d\theta} = - \frac{C_D A_{proj} D \rho_f}{2m_p} (\tilde{u}_y - \tilde{v}_y) \frac{(|\vec{u} - \vec{v}|)}{v_o} \qquad (21)$$

It is convenient to introduce the particle Reynold's number based upon relative velocity defined as:

$$Re_p = \frac{(|\vec{u} - \vec{v}|) d_p}{\nu_f}$$

Carrying the development forward just for the x-component, (21) may be transformed into

$$\frac{du_x}{d\theta} = - \frac{C_D Re_p}{2} \left(\frac{A_{proj}}{m_p}\right) \frac{D\mu}{v_o d_p} (\tilde{u}_x - \tilde{v}_x)$$

For spherical particles $A_{proj}/m_p = 3/2\rho_p d_p$ and the inertial parameter Stk (Stokes number) as defined in equation (40) Chapter 4, automatically arises:

$$Stk = \frac{C\rho_p d_p^2}{18\mu} \frac{v_o}{D}$$

$$\frac{d\tilde{u}_x}{d\theta} = -\frac{C_D Re_p}{24\ Stk}(\tilde{u}_x - \tilde{v}_x)$$

or

$$\frac{24\ Stk}{C_D Re_p}\ \frac{d\tilde{u}_x}{d\theta} + \tilde{u}_x - \tilde{v}_x = 0 \tag{22}$$

Since the objective here is to find the particle trajectory, i.e., to find the position of the particle (x, y) as a function of time, it necessary to substitute $\tilde{u}_x = d\tilde{x}/d\theta$ and $d\tilde{u}_x/d\theta = d^2\tilde{x}/d\theta^2$. This gives

$$\frac{24\ Stk}{C_D Re_p}\frac{d^2\tilde{x}}{d\theta^2} + \frac{d\tilde{x}}{d\theta} - \tilde{v}_x = 0$$

and

$$\frac{24\ Stk}{C_D\ Re_p}\frac{d^2\tilde{y}}{d\theta^2} + \frac{d\tilde{y}}{d\theta} - \tilde{v}_y = 0 \tag{23}$$

as the equations which must be solved in order to get the trajectory. In the solution the inertial parameter Stk remains constant, but Re_p (and C_D with it, see Chapter 4) varies along the path with the relative velocity scalar $|\tilde{u} - \tilde{v}|$. For Stokes law particles, where $C_D = 24/Re$, (23) reduces to

$$\begin{aligned} Stk\frac{d^2\tilde{x}}{d\theta^2} + \frac{d\tilde{x}}{d\theta} - \tilde{v}_x = 0 \\ Stk\frac{d^2\tilde{y}}{d\theta^2} + \frac{d\tilde{y}}{d\theta} - \tilde{v}_y = 0 \end{aligned} \tag{24}$$

and the value of Stk must include the Cunningham factor as needed.

Note that here again is the problem of determining the appropriate drag coefficient for an accelerating particle. It is similar to the problem discussed in Chapter 4, I.B.1, c.3. The usual practice here, as recommended there, is to use the values of C_D obtained for steady motion, assuming that little error will thus be introduced because most applications will be for low values of Re_p.

To solve these equations in any given case, for example as illustrated in Fig. 3 of this Chapter, will require a knowledge of the pattern of fluid streamlines because $\tilde{v}_x$ and $\tilde{v}_y$ are in themselves functions of x and y, although they may be taken as independent of time. It is well-known that such a pattern of streamlines varies with the nature of the fluid regime as characterized by the Reynold's Number for the target: $Re_D = v_o D/\nu_f$. Hence a solution of (24) can be worked out only for specified type of target shape and for a fixed value of Re_D. A particular solution for a particle trajectory starting at a point (x_o, y_o) sufficiently upstream so that the curvature represented by D has no effect (i.e., at which $\tilde{u}_x = \tilde{v}_x = 1$, and $\tilde{u}_y = \tilde{v}_y = 0$), will in general be a function of Stk and Re_D. If the pattern is taken to be that of potential flow, i.e., $Re \to \infty$, then Re_D will not enter specifically and the solution will depend only upon Stk. It is evident why Stk is called the inertial parameter.

Solution of (23) or (24) is complex and usually can only be carried out by approximate numerical methods. Fuchs [1, page 136] and Tuttle [2] outline methods which may be employed. Solutions for several cases (see page 187) have appeared in the literature: for example flow about an infinitely long cylinder of diameter D, or about a sphere of diameter D. The flow regimes considered are at the two extremes of: (a) an ideal (inviscid) fluid, potential flow $Re_D \to \infty$, velocity components about a cylinder, in polar coordinates (r, θ; θ = 0 at upstream - centerline)

$$
\begin{aligned}
v_r &= - v_o\,(1 - D^2/4r^2)\cos\theta \\
v_\theta &= v_o\,(1 + D^2/4r^2)\sin\theta
\end{aligned}
\tag{25}
$$

Equations (25) are valid only on the upstream side of the cylinder. (b) A viscous fluid at low $Re_D < 1$, for a cylinder

$$v_r = C_L \left(1 - \frac{D^2}{4r^2} - 2 \ln \frac{D}{2r}\right) \cos \theta$$

$$v_\theta = - C_L \left(1 - \frac{D^2}{4r^2} + 2 \ln \frac{D}{2r}\right) \sin \theta \qquad (26)$$

$$C_L = \frac{v_o}{2 (2 - \ln Re_D)}$$

valid for $\frac{2r - D}{D} << 1$, i.e., close to the surface of the cylinder. Equation sets (25) and (26) are due to Lamb [3].

Corresponding equations are available for velocity components about a sphere. In polar coordinates (r, θ; $\theta = 0$ at centerline downstream) assuming polar symmetry:

(a) potential flow

$$v_r = v_o (1 - D^3/8r^3) \cos \theta$$

$$v_\theta = -v_o (1 + D^3/16r^3) \sin \theta \qquad (27a)$$

(b) viscous (Stokes) flow

$$v_r = v_o \left(1 - \frac{3D}{4r} + \frac{D^3}{16r^3}\right) \cos \theta$$

$$(27b)$$

$$v_\theta = -v_o \left(1 - \frac{3D}{8r} - \frac{D^3}{32r^3}\right) \sin\theta$$

For the purpose of modelling dust collection, such solutions are important insofar as they show whether a particle collides with a target or not. The long cylinder is taken as model for the fibers in a filter, and the sphere for the drops of liquid in a scrubber or for the granules in a granular bed filter. The results of the trajectory calculations must be displayed in terms expressing the proportion of particles in the stream which collide and are collected. This leads to the subject of aerodynamic capture to be discussed next.

II. AERODYNAMIC CAPTURE ON ISOLATED SINGLE TARGETS

A. Single Target Efficiency

The term aerodynamic capture is used to refer to the collision of particles with a target surface either due to inertia or to some other cause. It is customary to express the extent of this "capture" in terms of an "isolated single target efficiency," defined as that fraction of particles of a given size d_{p_i} in the stream which collide with the target surface. This may also be expressed in terms of the equivalent cross-sectional area of the gas stream from which particles are removed by collision with the target surface:

η_{T_i} = isolated single-target efficiency, particles of grade i

$$\eta_{T_i} = \frac{\text{cross-sectional area of gas stream cleared of particles by single target}}{\text{cross-sectional area of target projected upstream}} \tag{28}$$

For example, if the target be a long thin cylinder of diameter D, then there is a depth of stream b such that $\eta_{T_i} = b_i/D$, or in the case of a spherical target, a diameter of a stream tube b such that $\eta_{T_i} = (b_i/D)^2$. The determination of η_{T_i} will be considered first for an isolated single target, and later for a single target which is a member of an array of targets.

In addition to inertial impaction there are a number of other processes which may lead to collision and capture. Particles may be attracted to the target by electrostatic force caused by charges on the particles, on the collector, on both. Very small particles may be moved at random by Brownian diffusion, and so chance to hit the target if they pass close enough to it. Other forms of diffusion may be directional toward the target, as when a high temperature gradient exists near the target surface (thermophoresis), or when there is a concentration gradient of gas molecules (diffusiophoresis).

In some instances gravitational settling onto a target may be significant. The appropriate force, in each case, must be incorporated in equation (19).

Combinations of collecting forces may readily exist with each type present contributing its share to the total target efficiency. A single target efficiency for an isolated target may be assigned to each type of collecting force, e.g., η_{II} = isolated single-target efficiency due to inertial impaction, η_{BD} = isolated single-target efficiency due to Brownian diffusion, η_E = isolated single-target efficiency due to electrostatic charge, and so on. The total overall isolated single target efficiency will then be some combination of these, but not necessarily their sum. It is possible that a given particle might have been collected by any one of several mechanisms. If so, its collection should only be counted once in determining the overall target efficiency. This question is discussed further in G. below (see page 206).

Detailed methods of detecting which collecting mechanisms are operative in a given situation, and of determining their respective η values, are required. They are presented in the following sections. It will be found that for each mechanism certain dimensionless parameters may be identified which are uniquely associated with the magnitude of η. The calculation of these parameters will lead rather quickly to a determination of which collecting mechanisms are important in a given situation and which may be neglected. The characteristic parameter, or parameters, for the significant mechanisms may then be used to estimate the value of η for each one.

B. Capture by Inertial Impaction

Of all the possible aerodynamic capture mechanisms (especially for particles $d_p \geqq 1$ micron), that due to inertial impaction is undoubtedly the most common and has received the greatest amount of study. It is of basic importance in the collection of dust by fibrous filters, and by scrubbers. It is also very important in certain particulate sampling devices (such as impingers, Anderson samplers and the like) and in probes used for stack sampling.

To determine η_{II} for a given case it is necessary to solve the pair of differential equations as (23), or (24), subject to the existing flow field given by $v_x = v_x(x, y)$ and $v_y = v_y(x, y)$, and to find the limiting trajectory for collision to occur. This establishes the position of the distance b, as illustrated in Figure 5 for impaction on a cylinder or a sphere. Then η_{II} is calculated by (28). For a particular shape of target and flow field, the value of η_{II} will be found to be a function of Stk and Re_D in general, or merely of Stk in the case of potential flow.

Calculations of the type outlined have been made for quite a few different target shapes. Usually they have been done for the case of large Re_D where potential flow (inviscid) fluid may be assumed in order to get the v_x, v_y values. Ranz [5] made a survey of these in 1956. More recent summaries are given by Strauss [4], Golovin and Putnam [6], May and Clifford [7], and Tuttle [2] who conducted a comprehensive literature search.

As a typical example of this work, two figures dealing with a cylindrical target are reproduced from Ranz's report. Figure 6 shows η_{II} as a function of $\sqrt{Stk}$, at various values of $Re_p^2/2\ Stk = 9\rho_f^2 v_o/\rho_p\mu$, as determined from equations (23) for very high values of Re_D (potential flow) [8]. On this figure, the capture of Stoke's law particles, as determined by equations (24) is presented as the special case identified by the curve for $Re_p^2/2Stk = 0$. Figure 7 shows the effect

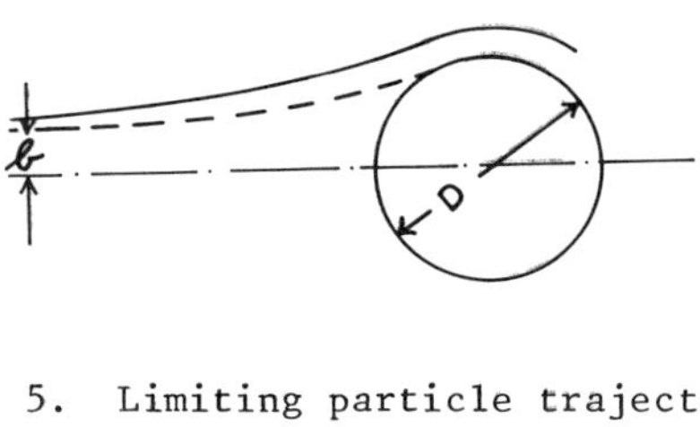

FIG. 5. Limiting particle trajectory.

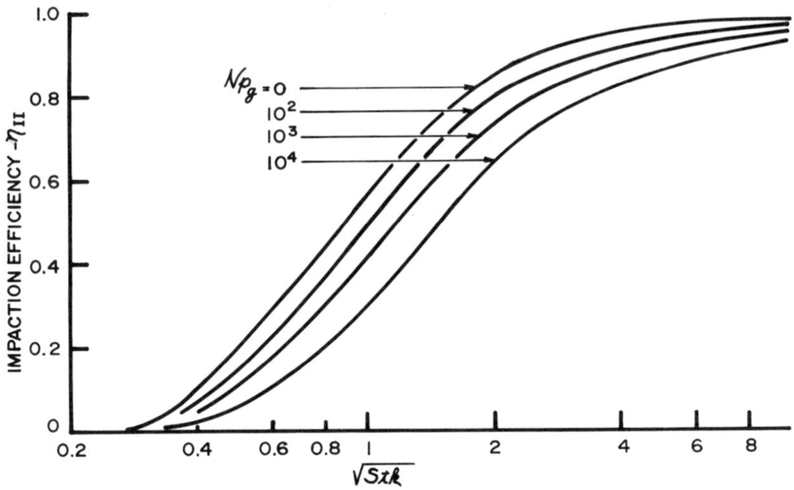

FIG 6. Capture efficiency by inertial impaction on spheres, potential flow. Reproduced by permission of Pennsylvania State University.

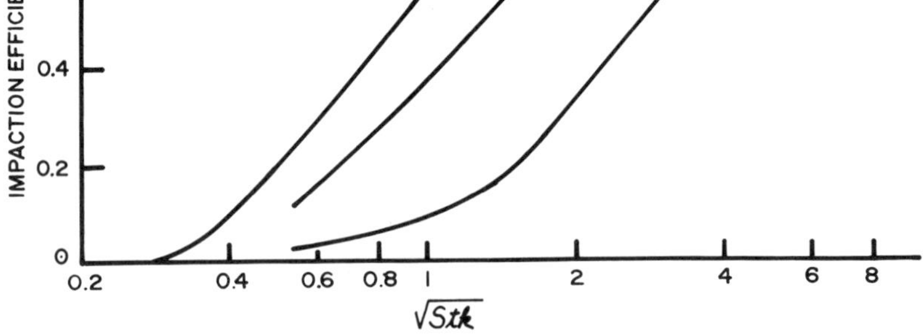

FIG. 7. Capture efficiency by inertial impaction on spheres, effect of flow regime. Reproduced by permission of Pennsylvania State University.

of the fluid flow pattern, as represented by different values for Re_D, all for Stoke's law particles only [5, 9, 10]. The top curve in both figures is the same.

On these figures $\sqrt{Stk}$ is used instead of Stk in order to display the particle size in dimensionless form as the independent variable. Thus

$$\sqrt{Stk} = \frac{d_p}{\left(\frac{18\mu D}{C\rho_p v_o}\right)^{1/2}} \text{, a dimensionless particle size.}$$

Bearing in mind, however, that Stk also is equal to x_s/D (see equation (40) Chapter 4), it is evident that as the stopping distance x_s approaches the same order of magnitude as the diameter of the target, the collection efficiency approaches the order of magnitude of 1.

Solutions for η_{II} in the form of equations or charts of the type of Fig. 6 or 7, and some experimental results, are available for at least the following shapes in the references listed:

Shape	Reference
Cylinder	[4], [5], [6], [9], [10], [17]
Sphere	[4], [5], [6], [11], [12], [13]
Round jet directed against flat plate	[5], [14], [15]
Rectangular jet directed against flat plate	[5], [9], [15], [16]
Disc collector	[5], [6], [7]
Ribbon collector	[5], [6], [7], [10]
Elbows	[5], [18], [19]
Wave-shaped channel	[5]
Recessed trough	[5], [6]
Rectangular half-body	[5], [6], [8]
Ellipsoids of revolution	[5], [6], [11]
Airfoils	[6]
Sampling tube (probe)	[20], [21]

Some of these are of primary interest in collection of particulate samples from flowing streams.

The S-shaped appearance of a number of these graphs has suggested that they might be linearized by replotting on log-probability coordinates in a manner similar to that for cumulative percent-finer-than particle size as discussed in Chapter 2. Such plotting does give reasonably straight lines for a number of η_{II} vs $\sqrt{Stk}$ charts available, particularly through the middle range of collection efficiencies. Whitby [22] presented a list of values for median Stk and for σ which may be used to reconstruct such plots. These are given in Table 1 and plotted on Figure 8.

Only a few equations are available to represent some of these results. For inertial impaction on cylinders at $Re_D = 10$, Pich [24] gives

$$\eta_{II} = \frac{Stk^3}{Stk^3 + 0.77\ Stk^2 + 0.22} \tag{29a}$$

For inertial impaction on spheres, the calculations of Herne [12] for potential flow have been fitted by Knettig and Beeckmans [56] to

$$\eta_{II} = 0.00376 - 0.464\ Stk + 9.68\ Stk^2 - 16.2\ Stk^3 \tag{29b}$$

where $0.0416 \leq Stk \leq 0.3$, and

$$\eta_{II} = \frac{Stk^2}{(Stk + 0.25)^2} \qquad Stk > 0.3 \tag{29c}$$

Cheng and Wang [18] give an elaborate equation for deposition in a 90° bend.

C. The Interception Effect

The calculation of limiting trajectories used to obtain values for η_{II}, makes no allowance for the size of the particle. When the particle is relatively large compared to the size of the collector, collection may be enhanced by direct interception. Fuchs [1, pg. 164] has shown how the increase may be estimated for two extreme cases: (a) the inertia of the particle is so large that it moves in a

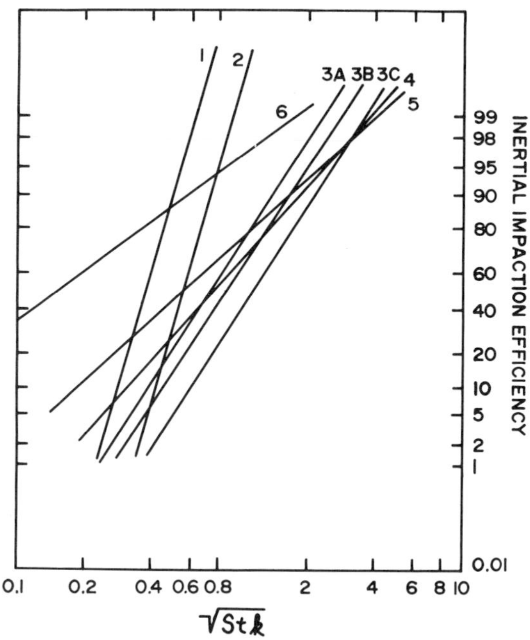

FIG. 8. Capture efficiency by inertial impaction on various shapes. 1. Jet on round plate. 2. Jet on rectangular plate. 3. Cylinder: a. $Re_D = 150$, b. $Re_D = 10$, c. $Re_D = 0.2$. 4. Sphere. 5. Rectangular half-body. 6. Focusing away. (See Table 1)

straight line, i.e., $Stk \to \infty$, (b) the particle possesses no mass, and moves in a streamline, i.e., $Stk = 0$.

In the first case, all particles within a stream tube of width D will collide with the collector. In addition particles passing within a distance $d_p/2$ will touch the collector, thus adding an increment of collection efficiency equal to η_{DI}: $\eta_{DI} = (d_p/D)$ for a cylinder, and $\eta_{DI} = (d_p/D)^2$ for a sphere.

In the second case, if the stream line on which the particle center is travelling approaches closer than $d_p/2$ to the collector, the particle will touch the collector. The increment of collection efficiency will then be, letting $R = d_p/D$ represent the basic parameter: for potential flow around a cylinder (using equations 25):

$$\eta_{DI} = 1 + R - 1/(1 + R) \simeq 2R \text{ (for } R < 0.1) \tag{30}$$

Table 1
Inertial Impaction Parameters
For Various Collector Target Shapes [22]

Collector	$\sqrt{Stk_M}$	σ_g
Jet on plate, round	0.373	1.24
Jet on plate, rectangular	0.55	1.24
Cylinder	$1.253 Re_D^{-0.0685}$	1.65
	$0.2 < Re_D < 150$	
Sphere	0.77	1.91
Ribbon	0.54	2.22
Trough or cup (Impinger)	0.46	1.91
Rectangular half body	0.54	2.32
Sweeping Bend	0.71	1.80
Focusing away	0.14	2.9

for potential flow round a sphere (using equations 27):

$$\eta_{DI} = (1 + R)^2 - 1/(1 + R) \simeq 3R \text{ (for } R < 0.1) \tag{31}$$

for viscous flow round a cylinder (using equations 26):

$$\eta_{DI} = \frac{1}{2.002 - \ln Re_D}\left[(1 + R)\ln(1 + R) - \frac{R(2 + R)}{2(1 + R)}\right]$$
$$\simeq \frac{R^2}{2.002 - \ln Re_D} \text{ (for } R < 0.07, \text{ and } Re_D << 0.5) \tag{32}$$

for viscous flow round a sphere:

$$\eta_{DI} = (1 + R)^2 - 3(1 + R)/2 + 1/w(1 + R) \simeq 3R^2/2$$
$$\text{(for } R < 0.1) \tag{33}$$

D. Capture by Flux Forces

In Chapter 4 the movement of particles due to various forces of bombardment by gas molecules was discussed. This includes Brownian diffusion, thermophoresis, and diffusiophoresis. Under appropriate conditions each of these may lead to capture of particles on targets. Collectively these kinds of capture are said to occur "by flux forces" $\vec{F}_F$ They are only significant for very small particles, $d_p \ll 1$ micron, for which inertial effects are negligible.

Capture by flux forces may be taken into account by adding an external force term, for each flux force acting, to the righthand side of Newton's Law. The flux forces may be converted into corresponding dimensionless flux velocities $\vec{u}_F$, to be added to equations (23) or (24). Since flux forces are only significant in the range of Stokes law particles, the relationship for conversion is:

$$\vec{F}_F = 3\pi\mu d_p u_F/C \tag{34}$$

and the dimensionless velocity term is defined as:

$$N_F = \frac{u_F}{v_o} = \frac{u_B + u_T + (u_D + u_{St})}{v_o} \tag{35}$$

called "flux deposition number" by Calvert, et. al., [23]. Equations (24), for very small values of Stk, become

$$\begin{aligned} \frac{d\tilde{x}}{d\theta} - \tilde{v}_x - \tilde{u}_{F_x} &= 0 \\ \frac{d\tilde{y}}{d\theta} - \tilde{v}_y - \tilde{u}_{F_y} &= 0 \end{aligned} \tag{36}$$

Aerodynamic capture efficiency by flux forces may be calculated by solving these equations in a manner similar to that used for inertial impaction, incorporating an appropriate fluid velocity pattern. This has only been done for several special cases of interest in particle collection. Brownian diffusion (u_B) may arise in any system, but thermophoresis (u_T) will occur only when there is a large temperature gradient maintained between gas and target, and diffusiophoresis (u_D) only when the target is a liquid surface, or where condensation of vapor may occur on a target surface.

1. Brownian Diffusion

A small particle, having little mass and therefore little inertia, will tend to follow the fluid streamlines in the neighborhood of a target. Bombardment by gas molecules, setting the particle into Brownian motion, will cause it to wobble randomly about a streamline. If it passes close enough to a target surface it may thus collide with that surface.

The analogy between Brownian diffusion and true diffusion in gases has been presented in terms of equation (56), Chapter 4. This equation may be placed in dimensionless form by using the same set of reference variables as for equations (20), yielding:

$$\frac{d\tilde{n}}{d\theta} = \frac{\mathcal{D}}{v_o D}\left(\frac{\partial^2\tilde{n}}{\partial x^2} + \frac{\partial^2\tilde{n}}{\partial y^2} + \frac{\partial^2\tilde{n}}{\partial z^2}\right)$$

where $\tilde{n} = n/n_o$. A dimensionless parameter automatically arises, which is called the Peclet number:

$$\mathrm{Pe} = \frac{v_o D}{\mathcal{D}}$$

This is a measure of the amount of convective transport relative to the diffusive transport of the particles. The larger the value of Pe, the less important is the diffusional process. A related dimensionless group, called the Schmidt number, may be formed by

$$\mathrm{Sc} = \mathrm{Pe}/\mathrm{Re}_D = v_o D\mu/\mathcal{D}\, v_o \rho_f D = \nu_f/\mathcal{D} \tag{37}$$

In these numbers, the value of $\mathcal{D}$ is obtained by the methods discussed in Chapter 4, II. B. 1. A Brownian flux velocity u_B has been defined as

$$u_B = \mathcal{D}/\delta_B \tag{38}$$

where δ_B is the boundary layer thickness of the fluid over the target surface.

Using these concepts, methods of calculating the single target collection efficiency by Brownian diffusion, η_{BD}, have been developed by several authors. The basic assumption is made that during the time the gas stream flows past the target, the particles from a certain layer are able to diffuse to the target surface. The thickness of this layer is proportional to $(\mathcal{D}\cdot t)^{1/2}$. The appropriate length of time to use depends upon the fluid velocity distribution. Thus in the viscous flow regime $\eta_{BD} = f\,(Pe, Re_D)$ and in the potential flow regime $\eta_{BD} = f(Pe)$. Good summaries of the results of various methods of calculation are given by Strauss [4] and by Pich [24].

a. Cylindrical Targets. For viscous flow, the original method is due to Langmuir [25] using the velocity field distribution given by Lamb, equations (26). His results may be expressed as

$$\eta_{BD} = \frac{1.71}{(2 - \ln Re_D)^{1/3}} Pe^{-2/3} \tag{39}$$

Similar equations have been derived by Natanson [26], Friedlander [27], and Crawford [28] in which the constant factors are 2.92, 2.22 and 2.854 respectively. Likewise an equation by Torgeson [29], differs but slightly in numerical values:

$$\eta_{BD} = \frac{2.06}{(2 - \ln Re)^{0.4}} Pe^{-0.6} \tag{40}$$

Ranz [30], using the analogy between heat and mass transfer, gives

$$\eta_{BD} = \frac{\pi}{Pe}\left(\frac{1}{\pi} + 0.55\ Re_D^{\ 1/2}\ Sc^{1/3}\right)$$

$$\eta_{BD} = \frac{1}{Pe} + 1.727\ \frac{Re_D^{\ 1/6}}{Pe^{2/3}} \qquad (41)$$

which is recommended for $0.1 < Re_D < 10^4$, and $Sc < 100$ $(10 < Pe < 10^6)$.

For potential flow, where the velocity field is independent of Re and corresponds to conditions of high Re, Stairmand [31] and Natanson [26] have presented similar results

$$\eta_{BD} = K/Pe^{1/2} \qquad (42)$$

where $K = \sqrt{8} = 2.83$ (Stairmand), or $K = \sqrt{32/\pi} = 3.19$ (Natanson).

From an inspection of these formulas, it is clear that collection efficiency due to Brownian diffusion will be very small unless the Peclet number is small, say less than 10. It is larger at higher Reynold's numbers (based upon target size). Note that it is theoretically possible to have a value of $\eta_{BD} > 1$ because Brownian diffusion may cause particles from beyond the distance D to collide with the target.

Example 2. Compare the relative importance of aerodynamic capture by inertial impaction, direct interception and Brownian diffusion for particles of unit density, having d_p ranging from 0.001 μm to 20 μm, upon a cylinder of 100 μm diameter from dry air at 20°C and 1 atm flowing at 10 cm/s.

$$\text{The value of } Re_D = \frac{10 \times (100 \times 10^{-4}) \times 1.205 \times 10^{-3}}{1.81 \times 10^{-4}} = 0.66$$

so the viscous flow velocity field must be used. Selecting the appropriate equations and charts, as listed, the various parameters and corresponding capture efficiencies may be determined as tabulated below.

d_{pi}-μm	Stk	η_{II}-%	R	η_{DI}-%	Pe	η_{BD}-%
0.001	-	-	-	-	2.63	66.7
0.01	-	-	-	-	2.50×10^2	3.2
0.2	-	-	-	-	1.64×10^4	0.2
1	3.54×10^{-3}	0	0.01	0	3.70×10^5	0.02
10	3.08×10^{-1}	3(9.1)	0.1	0.4	-	-
20	1.23	11(57.3)	0.2	1.5	-	-

η_{II} estimated from Fig. 7; values in ( ) from equation (29a), η_{DI} calculated from equation (32), η_{BD} calculated from equation (39).

It is evident from this example that diffusion contributes nothing as a collection mechanism unless the particles are very small. But as the particle size decreases, collection by diffusion increases while that due to inertial impaction decreases. At some intermediate particle size (e.g., 1-micron, in the example) there is a minimum collection where neither inertia nor diffusion is effective.

A similar effect is obtained with respect to velocity v_o. At low velocities diffusion will predominate, at high velocities inertia will be predominant. There will be a minimum collection efficiency determined primarily by the size ratio R (direct interception), at some intermediate velocity.

b. Other Target Shapes. Except for spheres very little has been done to estimate collection by Brownian diffusion on shapes other than cylindrical.

For isolated spherical targets the following are available. Johnstone and Roberts [32] suggest (D = diameter of sphere):

$$\eta_{BD} = \frac{4}{Pe}(2 + 0.557\, Re_D^{1/2}\, Sc^{3/8}) \tag{43}$$

or

$$\eta_{BD} = \frac{8}{Pe} + 2.23\, Re_D^{1/8}\, Pe^{-5/8}$$

Crawford [28, pg.380] derives a formula:

$$\eta_{BD} = \frac{4.18}{Sc^{2/3} Re_D^{1/2}} = 4.18\, Re_D^{1/6}\, Pe^{-2/3} \tag{44}$$

while Pfeffer [33] gives:

$$\eta_{BD} = 3.96\ Pe^{-2/3} \tag{45}$$

Strauss [4, pg. 291] cites:

$$\eta_{BD} = 8\sqrt{2/Pe}/3\pi = 1.20\ Pe^{-1/2} \tag{46}$$

Strauss [4] also gives, for a side of a strip of width W:

$$\eta_{BD} = \sqrt{8\mathcal{D}/\pi v_o W} \tag{47}$$

2. Thermophoresis

Whenever a target surface is sufficiently colder than the gas, so that a large temperature gradient exists across the fluid boundary layer on the target, capture will occur due to thermophoresis. The thermal flux force F_T is given by equation (56) in Chapter 4, and the thermal velocity u_T by equation (55). These will always act normal to the target and the velocity components $\tilde{u}_{T_x}$ and $\tilde{u}_{T_y}$ to be used in equations (35) must be reckoned accordingly. A thermal flux deposition number may be calculated using (34). This will obviously be a function of the dimensionless group Th given by equation (55), Chapter 4.

The flux force deposition number for very small particles (Th≃0.50) may be estimated

$$N_{FT} = \tilde{u}_T = -\frac{0.50}{v_o}\frac{\nu_f}{T}\frac{dT}{dx} \tag{48}$$

This may be appreciable even at small values of ΔT between target surface and gas because the gradient will be across a very thin thermal boundary layer, say of the order of 10^{-3} cm thick. In dry air at 1 atmos., 300°C, with ΔT = 10°C, and v_o - 10 cm/sec $u_T = 4.2 \times 10^{-4}$ (-10×10^3) = 4.2 cm/sec, and N_{FT} = 0.42.

To make this calculation it is necessary to have a method of estimating the thermal boundary layer thickness. For spherical targets, according to Johnstone and Roberts [32] this will be

$$\Delta x = D/\left(2 + 0.557\,Re_D^{\ 0.5}\,Pr^{0.375}\right) \qquad (49)$$

where $Pr = C_p\mu/k_f$ (C_p = heat capacity, k_f = thermal conductivity) and all gas properties are evaluated at the mean boundary layer temperature.

Thermophoresis is essentially independent of the particle size. For very small particles $Kn \gg 1$, $Th \simeq 0.50$ as taken above. However Th is a weak function of particle size, as discussed in Chapter 4. For estimates of aerodynamic capture where u_T is likely to be significant, however, Th may be taken as constant.

Example 3. No generalized calculations of capture efficiency due to thermophoresis have been found in the literature. However Pilat and Prem [34] have worked out some specific sample cases using the methods outlined above for a water drop 100 μm as a spherical target moving at 30 cm/s. Holding the air temperature constant at 65°C they took the drop temperature to range from 10°C to 82°C, and calculated isolated target efficiencies for particles ranging from d_{p_i} = 0.01 μm. In this range the value of u_T is essentially independent of d_p. Some sample results are:

T_{drop} -°C	Δx - cm	dT/dx - °c/cm	u_T-cm/sec
10	-3.723×10^{-3}	-1.477×10^{4}	3.96
48	-3.726×10^{-3}	-4.562×10^{3}	1.29
65	-3.724×10^{-3}	0	0
82	-3.719×10^{-3}	$+4.571 \times 10^{3}$	-1.35

Δx calculated from (49), $dT/dx \simeq (T_{air} - T_{drop})/\Delta x$, u_T calculated from (48), using Th = 0.50

Unfortunately it is impossible to obtain the values of η_{Th} from the published work, because they are incorporated into an overall efficiency. It is evident however that they have an appreciable effect at drop temperatures away from 65°C.

3. Diffusiophoresis

Whenever a gaseous concentration gradient exists, diffusiophoresis would enter into the capture process. It will aid in capture if the gradient is toward the target, and oppose capture if the gradient is away from the target. The most likely circumstance under which this could arise would be in case the target is a liquid surface. Condensation of vapor on the liquid would contribute to capture, evaporation of liquid would oppose capture. Here the Stefan flow discussed in Chapter 4, section II.B, will come into play. The diffusion force F_D may be calculated by equation (66), Chapter 4, the diffusion velocity $(u_D + u_{St})$ by equation (65).

The flux force deposition number will be

$$N_{FD} = \tilde{u}_D = -\left(\frac{\sqrt{m_B}}{n_A\sqrt{m_A} + n_B\sqrt{m_B}}\right)\frac{D_{AB}}{v_o n_A}\ \frac{\partial n_B}{\partial x} \tag{50}$$

It is evident that N_{FD} is completely independent of particle size. In case the diffusion effect is created by a liquid surface, the values of n_A and n_B may be represented in terms of the partial pressure of the vapor in a given system and (50) becomes

$$N_{FD} = -\frac{K}{v_o}\ \frac{dp}{dx} \tag{51}$$

so that $N_{FD} > 0$ for condensation $dp/dx < 0$ and $N_{FD} < 0$ for evaporation $dp/dx > 0$ and K is a constant depending upon the system, the temperature, and the total pressure. At the liquid surface p will be equal to the saturation vapor pressure at the temperature of the liquid. For the system water-vapor diffusing in air at 0°C, 1 atm. this becomes

$$N_{FD} = -\frac{1.9 \times 10^{-4}}{v_o}\ \frac{dp_w}{dx}$$

where dp_w is expressed in mb/cm.

A relationship similar to (49) is available [32] for estimating the boundary layer thickness Δx due to diffusivity in the case of a spherical target

$$\Delta x = D/ (2 + 0.557\ Re_D^{0.5}\ Sc_D^{0.375}) \tag{52}$$

Where the Schmidt number is based upon the diffusivity of the target vapor, instead of the particles as in (37).

Sparks and Pilat [35] performed calculations for the combined effect of inertia and diffusion flux force, using equations (24) with $u_D = N_{FD}$ added, such as is illustrated in (36). They assumed potential flow around a water drop, using (27) for the velocity components, and calculated η as a function of Stk for three cases: $dp/dx = -10^5$mb/cm, $dp/dx = 0$, and $dp/dx = 10^5$mb/cm, all with $v_o = 100$ cm/s. The results are presented graphically, since the equations can only be solved numerically. At Stk = 1, for example, $\eta_{II} = 0.45$ whereas with condensation the combined efficiency rises to 0.95 and with evaporation it drops to 0.16.

Example 4. Prem and Pilat [34] also incorporated diffusiophoresis into their sample calculations mentioned above under thermophoresis, for the same set of conditions. The air was taken to be saturated with water vapor at 65°C. For drop temperature < 65°C, condensation occurs $dp/dx < 0$, and diffusiophoresis augments thermophoresis as a combined flux force capture mechanism. For drop temperature > 65°C the reverse is true. Some samples of diffusion velocities calculated are:

T_{drop} - °C	p_v^{mmHg}	Δx - cm	dp/dx-mb/cm	$(u_D + u_{St})$ - cm/sec
10	9.21	3.77×10^{-3}	-6.30×10^4	11.9
48	83.7	3.76	-3.68×10^4	7.0
65	187.5	3.72	0	0
82	384.9	3.72	7.06×10^4	-13.4

Δx is calculated from (52), p_v is vapor pressure of water at indicated temperature $(u_D + u_{St})$ is calculated from (51), and is independent of d_p.

Again it is not possible to separate out the diffusiophoretic effect on the capture efficiency, but it is evident that it is much greater than the thermophoretic effect. To give some idea of the combined influence it may be noted that for 1 μm particles the total isolated drop capture efficiency at 65°C is given as 0.015, while at 10°C it is 2.7 (270%!) and at 82°C it is below 0.001. Flux force capture obviously may be very significant.

E. Collection by Electrostatic Attraction

Some aerosol particles seem to acquire electrostatic charges rather easily either during their generation or during flow through a gas stream. Likewise target elements such as the fibers making up a filter, may acquire charges due to the friction caused by the passage of a gas stream over them. Sometimes this effect can be enhanced by impregnating the target fibers with resinous material. Furthermore targets may be charged by imposing an electrostatic field on the collector. Whenever any of these situations occur, electrostatic forces are set up which will influence the particle collection process.

Unless an electrostatic field is deliberately maintained, the charge on the target elements will be unstable and will tend to decrease with time, owing to such factors as conductivity of the target material, passage of ionized gas molecules, radioactive irradiation, deposition of charged particles, and humidity (which tends to ground the target).

In the absence of an external electric field, there are three general cases to be considered: (i) Charged particles-neutral target. The charge on the particles induces an image charge, opposite in sign, on the collector which results in a force of attraction between particle and target, called F_{EM}. (ii) Charged target-neutral particles. An image charge of opposite sign is induced on the particles, generating a force of attraction, F_{EI}. (iii) Charged particles - charged target. Coulombic forces of attraction or repulsion act, depending upon whether the particles and the target have unlike or like charges. This force is called F_{EC}. In addition, wherever the particles are charged in the same sense, a repulsion force is produced among themselves. This is called the space-charge effect F_{ES}. Table 2 lists the calculation of these forces both for isolated spherical and isolated cylindrical shaped targets, provided the charge on the collector is constant.

The analysis of the effect of these forces, together with the other mechanisms, upon aerodynamic capture is complicated. In principle the appropriate electrostatic forces should be summed:

$$F_E = F_{EM} + F_{EC} + F_{ES} \tag{53}$$

where there is a constant charge on the collector. The total electrostatic force is then added to all other forces in Newton's Law for the calculation of trajectories. As in the case of inertial impaction, it is necessary to assume some fluid velocity field for the calculations. The resulting differential equations would be similar to (19) through (24), but with an added term including F_E in the sum of forces on the right side.

There will correspond to each such force term a dimensionless parameter which is formed for Stokes law particles (including Cunningham correction) when the equations are put into the dimensionless form of (24). The collection efficiency η_E for an isolated single target is then found to be a function of Re_D, Stk, and the corresponding electrostatic parameters, which are listed in Table 3. Inertialess particles may be considered by placing Stk = 0.

Kraemer and Johnstone [36] have calculated values of η_E for inertialess particles on isolated spherical targets, allowing for direct interception. Their results are given in the form of charts giving η_E as a function of K's in each of the three basic cases, calculated for potential flow and viscous flow, each at several values of R. These charts are also reproduced by Strauss [4, pg. 304].

Table 2

Electrostatic Forces Between Spherical Particle and Collector of Constant Charge

Nature of Force	Spherical Collector [36]	Cylindrical Collector [37]
	D = diameter Q = charge on collector (constant)	D = diameter ε_D = dielectric constant of collector
F_{EM} - Image: charged particle-neutral collector	$\frac{q^2 D}{8\pi\varepsilon_o r^3} - \frac{2q^2 Dr}{\pi\varepsilon_o(4r^2 - D^2)^2}$	$\frac{q^2}{4\varepsilon_o(r - \frac{D^2}{2}} \frac{\varepsilon_D - 1}{\varepsilon_D + 1}$
F_{EI} - Image: neutral particle-charged collector	$-\left(\frac{\varepsilon - 1}{\varepsilon + 2}\right) \frac{d_p{}^3 Q^2}{16\pi\varepsilon_o r^5}$	$\frac{4Q^2}{\varepsilon_o} \left(\frac{\varepsilon - 1}{\varepsilon + 2}\right) \frac{\left(\frac{dp}{2}\right)^3}{r^3}$
F_{EC} - Coulombic: charged particle-charged collector	$\frac{Qq}{4\pi\varepsilon_o r^2}$	$\frac{2Qq}{\varepsilon_o r}$
F_{ES} - Space-charge: repulsion between particles	$\frac{-q\ D\ N}{24\varepsilon_o r^2}$	

N = number of particles per unit volume
r = distance between particle and collector
q = charge on particle
ε = dielectric constant of particle
ε_o = specific inductive capacity of space = 8.85×10^{-21} coulombs^2/dyne cm^2

Table 3
Dimensionless Parameters K Representing Electrostatic Forces on Stokes Law Particle

Nature of Force	Spherical Collector [36, 57] Qa = charge per unit area of collector		Cylindrical Collector [37] Q = charge per unit length of collector	
	Parameter - K	Efficiency - η_E	Parameter - K	Efficiency - η_E
F_{EM} - charged particle; neutral collector $K = K_{EM}$	$\frac{Cq^2}{3\pi^2 \mu d_p v_o \varepsilon_o D^2}$	$\simeq 1.58\, K_{EM}^{1/2}$ (b) [57] $\simeq 2.89\, K_{EM}^{0.353}$ (a) [57] $0.002 \leq K_{EM} \leq 0.1$	$\frac{(\varepsilon_D - 1)}{(\varepsilon_D + 1)} \frac{Cq^2}{3\pi \mu d_p D^2 v_o \varepsilon_o}$	$(6\pi)^{1/3} K_{EM}^{1/3}$ (a) $\frac{2}{(2-\ln Re)^{1/2}} K_{EM}^{1/2}$ (b)
F_{EI} - neutral particle, charged collector $K = K_{EI}$	$\left(\frac{E-1}{E+2}\right) \frac{2Cd_p^{2} Q_a^{2}}{3\pi \mu D v_o \varepsilon_o}$	$\simeq \left(\frac{15\pi}{8} K_{EI}\right)^{0.4}$	$\frac{4}{3\pi}\left(\frac{\varepsilon-1}{\varepsilon+2}\right) \frac{Cd_p^{2} Q^2}{D^3 \mu v_o \varepsilon_o}$	$\left(\frac{3\pi}{2}\right)^{1/3} K_{EI}^{1/3}$ $\frac{2r}{D} >> 1$
F_{EC} - charged particle, charged collector $J = K_{EC}$	$\frac{CqQ_a}{3\pi \mu d_p v_o \varepsilon_o}$	$-4K_{EC}$ (a) and (b)	$\frac{4QqC}{3\pi \mu d_p D v_o \varepsilon_o}$	$-\pi K_{EC}$
F_{ES} - repulsion bet. particles	$\frac{Cq^2 DN}{18\pi \varepsilon_o \mu d_p v_o}$			(a) potential flow (b) viscous flow

C = Cunningham correction factor
η_E = isolated single-target efficiency (may be > 1)
Qa, Q,q may be expressed as integral multiples of e
e = unit electronic charge = 1.601×10^{-19} coulombs

In general they show log η roughly linear in log K, with higher efficiencies for potential than for viscous flow. Values of $\eta_E > 1$ are of course possible.

Nielsen and Hill [57] have extended this study considerably, using the same three basic parameters as Kraemer and Johnstone and adding two more. These deal with the presence of an external electric field, one where particles only are charged and the other where neither particles nor collector are charged. Their results agree with Kraemer and Johnstone in the duplicated cases, and their chart is similar, producing some useful approximations as tabulated.

Pich [24] has stated the corresponding parameters (no external field) for the case of cylindrical collectors, and has quoted the expressions derived by Natanson [37] for the collection efficiencies. These are also listed in Table 3.

The K parameters given in Table 3 may be formed, as shown by Ranz and Wong [38], by evaluating the corresponding force at an appropriate value of r and then dividing by the drag force taken as $3\pi d_p \mu v_o / C$.

Example 5. Illustrate the effect of electrostatic charges by calculating the isolated target efficiencies for particles of unit density with d_p ranging from 0.001 to 1 microns, captured on a conducting cylinder 100 microns in diameter from a stream at 20°C, 1 atm, dry air flowing at 10 cm/s. Consider the particles to be charged with 10 electrons each, and with 300 electrons each.

As calculated in previous examples Re = 0.66 for the conditions, so η_E will be calculated from K_{EM} according to case (b) in Table 3. Since the value of ε_D is not given, take $(\varepsilon_D - 1)/(\varepsilon_D + 1) \simeq 1$ and calculate K_{EM} for each particle size from

$$K_{EM} = \left(\frac{C}{d_p}\right) \frac{[n(1.601 \times 10^{-19})]^2}{3\pi(1.81 \times 10^{-4})(100 \times 10^{-4}) \times 10 \times 8.86 \times 10^{-21}}$$

$$= \left(\frac{C}{d_p}\right) \times 1.70 \times 10^{-12}\ n^2$$

where n is the number of electronic charges per particle. Then

$$\eta_E = \frac{2}{(2 - \ln 0.66)^{1/2}} K_{EM}^{1/2} = 1.27\ K_{EM}^{1/2}$$

The results are tabulated:

d_p - μm	n = 10 electrons		n = 300 electrons	
	K_{EM}	η_E	K_{EM}	η_E
0.001	0.367	0.77	330	23.1
0.01	3.77×10^{-3}	0.078	3.40	2.3
0.1	4.5×10^{-5}	0.008	0.041	0.26
1	2×10^{-6}	0.002	1.77×10^{-3}	0.053

It is evident that the effect may be very significant on fine particles. Unfortunately we do not often have sufficient data (especially for n) to make such calculations. Efforts are being made to utilize this kind of effect to enhance the collection of very fine particles.

F. Capture by Gravity

Gravity may affect aerodynamic capture if conditions are such that the terminal settling velocity of particles is appreciable compared to the capture velocities which may be operating. From a horizontal slow moving gas stream, large particles may tend to settle onto a target from above. In a vertical gas stream flowing downward toward a target, gravity will enhance the capture, while if the stream is flowing upward gravity will tend to detract from the capture. This effect has been demonstrated conclusively by experiments such as those of Thomas and Yoder [39].

In Chapter 4, a dimensionless parameter G · Stk was developed and given in equation (41) to be

$$G \cdot Stk = u_s/v_o$$

Ranz and Wong [38] have shown that this would represent the isolated single target efficiency of capture by settling:

$$\eta_G = G \cdot Stk = Cgd_p{}^2\rho_p/18\mu v_o \qquad (54)$$

It is evident that for η_G to be significant d_p must be relatively large and v_o small. This situation does not arise frequently in the problems of capture of fine particles. In most collecting situations G · Stk will not be important.

G. Capture by Combinations of Forces

When aerodynamic capture is occuring due to the simultaneous action of more than one significant mechanism, the overall isolated target efficiency may be estimated by taking the net penetration to be the product of the penetrations resulting from each mechanism acting alone, e.g.:

$$P_T = P_{II} \cdot P_{DI} \cdot P_{BD} \cdot P_E \cdot \cdot \cdot \qquad (55)$$

This is equivalent to

$$\eta_T = 1 - (1 - \eta_{II})(1 - \eta_{DI})(1 - \eta_{BD})(1 - \eta_E) \cdot \cdot \cdot \qquad (56)$$

The overall efficiency is not equal to the summation of the efficiencies for each mechanism operating alone. A given particle might have been collected by means of more than one mechanism, but its collection may be counted only once.

Crawford [28, pg. 382] has shown that, in the case of only two mechanisms acting, (56) is equivalent to each mechanism acting upon the average number of particles not collected by the other effect.

Several specific combined effects have been studied individually for cylindrical and spherical targets as listed below. Presumably these results would be preferable to (56) in calculations where they apply. Equations (55) or (56) can be used in the absence of a specific case study, wherein the several η values would each be estimated from appropriate formulas as given in the preceding sections.

1. Cylindrical Targets

a. Inertial Impaction Plus Direct Interception. Davies [40] has calculated this combination for $Re_D = 0.2$ (viscous flow field) and the results are given by

$$\eta_{II + DI} = 0.16 \ [R + (0.50 + 0.8R) \ Stk - 0.1052 \ R \ Stk^2] \tag{57}$$

b. Brownian Diffusion Plus Direct Interception. Friedlander and Pasceri [41] fitted their diffusion equation plus an interception term to several sets of experimental data and obtained excellent agreement, for viscous flow, with

$$\eta_{BD + DI} = 6Sc^{-2/3} \ Re_D^{-1/2} + 3R^2 Pe^{1/2} \tag{58}$$

Billings [42] however has pointed out that in certain experiments solid particles were collected with much greater efficiency than (58) predicts, while liquid particle behavior agreed well with (58). Perhaps this is due to electrostatic or other undetected extraneous effects in these experiments.

c. Inertial Impaction Plus Direct Interception Plus Brownian Diffusion. Davies [40] suggested simply adding 1/Pe to Stk in equation (57) in order to account for all three effects. This would be valid only for $Re_D = 0.2$.

2. Spherical Targets

a. Inertial Impaction Plus Direct Interception Plus Gravity. George and Poehlein [43] have solved equations analogous to (24) in spherical polar coordinates including not only the inertial effects (Stk) but also direct interception (R), and gravity ($G_D \cdot$ Stk, equation (54)). Potential flow and Stokes law particles were assumed. Their results are presented in a series of graphs in which the total isolated single target (sphere) collection efficiency is plotted against appropriate parameter.

One graph shows η_T vs $\sqrt{Stk}$ for R = 0 and G_D = 0, for R = 0 and $G_D = 8.8 \times 10^{-2}$, and for R = 0.1 and G_D = 0. Both target particle and collected particle were taken to be falling at their respective terminal velocities in a stationary gas. Consequently where $G_D > 0$, the total capture efficiency is reduced. Note that G_D is based upon target diameter, while Stk is based upon particle diameter. The results agree well with some previous calculations by Langmuir and Blodgett [44] and some experimental work by Jarman [45].

b. Inertial Impaction Plus Electrostatic Collection. Nielsen and Hill extended their work on electrical forces to include inertial impaction [58], producing an extensive calculation of particle trajectories and capture efficiencies for both potential and viscous flows. The results reveal a very interesting, and fairly complex, interplay of electrical and inertial forces. For example, with a strong coulombic force ($K_{EC} = -10.0$) and high inertia (Stk = 2.5) particles may pass completely around to the downstream side of the collector and cross the center-line several times in reverse zig-zag fashion before finally being collected on the downstream side. Over half of the total collection may take place on the downstream half ("back side") of the collector if $K_{EC} < -1.0$ and Stk is not very large. As Stk increases from zero, collection efficiency (initially > 1 for $K_{EC} < 0$) decreases, passes through a minimum and eventually increases again. Similar effects are observed in the case of the image forces (represented by K_{EM} and K_{EI}), and for either potential or viscous flow.

It is difficult to summarize briefly the results of all these cases, and impossible to present them quantitatively. One must study the collection of some thirteen charts presented. It may be useful to quote, in part, the authors' conclusions

> . . . under favorable conditions all of the electrical forces considered can significantly enhance particle collection over that obtainable by inertial impaction alone. . . . in many cases involving strong attractive electrical forces the collection efficiency is greatest for negligible particle inertia and decreases with increasing particle inertia. . . . the introduction of particle inertia to a given electrical force situation produces a greater change in η for potential flow than for Stokes (viscous) flow. With weak electrical forces η often passes through a minimum as Stk increases from zero. When a strong attractive radial force is present, a substantial portion of the total collection may occur on the downstream half of the collector.

One of the practical applications for this kind of calculation lies in the effort to improve the collection efficiency of scrubbers or granular beds by electrification of targets and/or particles. These applications will be made in the respective chapters. Nielsen and Hill point out that in the case of electrified wet scrubbers the operation should be at low relative droplet-to-gas velocities. Since $\mathrm{Stk} \propto v_o$ and any of the K_E's $\propto v_o^{-1}$, small relative velocities will decrease the effect of inertia and increase the effect of the electrical force.

c. Inertial Impaction Plus Flux Forces. Sparks and Pilat [35] have calculated the effect of diffusiophoresis upon inertial impaction for a target taken to be a spherical drop of water. Potential flow was assumed using equations (27) for the fluid velocity profile. For water-air at 0°C and 1 atm, equation (66), Chapter 4 becomes

$$(u_D + U_{St}) = -1.9 \times 10^{-4} \frac{dp}{dx}$$

where p is the partial pressure of the water vapor in millibars, x is the radial distance from the surface of the drop in centimeters, and v is in cm/s. Isolated single sphere aerodynamic capture efficiencies were calculated by using the equation (34) in combination with (24), and expressed as a function of the inertial impaction parameter K (here defined as K = 2 Stk) for three cases, all at v_o = 100 cm/s:

$\frac{dp}{dx} = 10^5$ mbar/cen — evaporation + inertial impaction $N_F = -0.19$

$\frac{dp}{dx} = 0$ — inertial impaction only $N_F = 0$

$\frac{dp}{dx} = -10^5$ mbar/cen — condensation + inertial impaction $N_F = 0.19$

Results are presented as a plot of $\eta_{II + D}$ vs K with a curve for each case. The effect of diffusiophoresis is very pronounced, for example at K = 1, η = 0.95 for condensation, η = 0.45 for impaction only, and η = 0.15 for evaporation. The calculations were continued to show the effect on the overall performance efficiency of a scrubber.

Pilat and Prem [34] extended these calculations to include the effects of Brownian diffusion and thermophoresis as well. This was for the specific case of a single water drop 100 μm diameter falling freely (u_s = 30 cm/sec) in air at 65°C, 100% relative humidity, 1 atm pressure. Although $Re_D \simeq 2$, potential flow pattern was assumed. The capture of particles ranging in diameter from 0.01 to 10 microns was calculated.

In order to show the effect of the several flux forces, the temperature of the drop was considered at 10°C 38°C, 49°C, 60°C, 65°C, and 82°C. At all drop temperatures below 65°C both thermophoresis and diffusiophoresis would aid in the capture, while above 65°C the reverse would be true. At 65°C only Brownian diffusion and inertial impaction would be acting. Brownian diffusion would contribute to particle capture at all temperatures, its effect increasing somewhat as the temperature increases. Temperature gradients ranged from

-1.48×10^4 °C/cm to 4.6×10^3 °C/cm, and water partial pressure gradients from -6.4×10^4 mbar/cm to 5.3×10^5 mbar/cm. These values are so large primarily because the boundary layer around the drop is so thin, of the order of 0.00037 cm.

The methods of calculation were similar to those employed by Sparks and Pilat. Flux force velocity terms were added to equations (24). These indicate that at a given temperature the flux velocity for Brownian diffusion increases markedly as particle size decreases, while those for thermophoresis and diffusiophoresis are essentially independent of particle size.

The interplay of the influence of the several mechanisms was shown clearly by a grade efficiency plot of η_{Total} (isolated sphere) vs particle size (0.01 - 10 μm). At 65°C (drop temperature) the efficiency has a minimum value of 0.003 at $d_p \simeq 0.09$ μm; Brownian diffusion causes an increase to 0.05 at $d_p = 0.01$ μm, while inertial impaction causes an increase to 1 at $d_p = 10$ μm. At 82°C, inertial impaction again brings about a high efficiency of 0.60 at $d_p = 10$ μm, but the adverse effect of thermophoresis and diffusiophoresis quickly predominates as particle size decreases, resulting in efficiency falling to 0.0001 at $d_p = 3$ μm. At all drop temperature below 65°C efficiencies are greater the lower the temperature, reaching about 2.5 at 10°C and being rather independent of particle size below about 1 μm. Thermophoresis clearly dominates, having values of $N_{FT} \simeq 0.12$ while those for diffusiophoresis are $N_{FD} \simeq 0.04$.

H. Capture Mechanisms and Parameters

Each of the capture mechanisms discussed above has been associated with a characteristic dimensionless group or parameter. The capture efficiency due to any one mechanism is determined by the corresponding parameter through formulas or charts. In a given set of conditions e.g., particle size, target geometry and size, fluid flow rate, temperature(s), charges present, etc, etc, a calculation of the parameters will reveal which mechanisms will be predominant. It will also show what the trend of capture efficiency will be corresponding to shifts in any of the variables. A summary of all parameters is given in Table 4.

1. Detection of Significant Mechanisms and Trends

Ranz and Wong [38] have stated a series of generalizations concerning mechanisms of collection, which are valuable guides. Some of them are quoted as follows

> 1. Whenever one parameter characterizing a mechanism of collection is much greater than all other parameters, collection can be attributed entirely to the mechanism which that parameter characterizes.
> 2. The efficiency of collection will be negligible for any given mechanism where the parameter characterizing that mechanism is less than 10^{-2} and will be of the order of unity when that parameter is of the order of unity.
> 3. Although the forces of the various mechanisms are additive, the resulting individual efficiencies are not directly additive. Each mechanism will contribute to a total efficiency, however, and no combination of favorable mechanisms will cause an efficiency lower than that expected for any one of the favorable mechanisms.
> 4. The efficiency of collection for a given mechanism will will be proportional to the size of the particle and the size of the collector (target), and to the stream velocity, in approximately the same manner that the parameter characterizing that mechanism is proportional to d_p, D, and v_o by definition.

These generalizations form the basis of practical use for the parameters in Table 4 as will be illustrated by some examples below. In addition to the trends mentioned in item 4, there are the important questions of the effects of the temperature, pressure, and composition of the gas stream. These influences may be determined also from an inspection of the parameters and will be discussed in the next section.

Emi and Yoshioka [46] have illustrated a very interesting graphical method of showing the interplay of the several mechanisms (II, DI, BD, and G) over a range of conditions. This consists of making a

Table 4

Summary of Collecting Mechanisms, Parameters, and Trends

Mechanism	Parameter(1)	Trends on $\eta^{(2)}$ d_p	D	v_o	Equations
All (effect of fluid velocity field)	$Re_D = \frac{v_o \rho_f D}{\mu_f}$	-	up	up	(25), (26), (27)
Inertial Impaction	$Stk = \frac{C\rho_p d_p^2 v_o}{18\mu_f D}$	up	down	up	(29), Fig. 7, 8, 9
Direct Interception	$R = \frac{d_p}{D}$	up	down	-	(30, (31), (32), (33)
Brownian Diffusion	$Pe^{-1} = \frac{\mathcal{D}}{v_o D}$	down	down	down	(39) thru (47)
Thermophoresis	$N_{FT} = \frac{u_T}{v_o}$	-	-	down	-
Diffusiophoresis	$N_{FD} = \frac{u_D + u_{St}}{v_o}$	-	-	down	-
Gravity	$G \cdot Stk = \frac{Cg\rho_p d_p^2}{18\mu_f v_o}$	up	-	down	(54)
Electrostatic	K_{EM} (See Table 3)	down	down	down	Table 3
	K_{EI} (See Table 3)	up	down	down	Table 3
	K_{EC} (See Table 3)	down	down	down	Table 3
	K_{ES} (See Table 3)	down	up	down	Table 3

Notes: (1) Increase in any parameter means increase in capture efficiency due to that mechanism.

(2) Trend means effect on η as variable is increased.

plot a coordinates of particle size vs. stream velocity and indicating the domains in which each mechanism is predominate. The exact appearance of such a plot will depend upon the sample conditions and precise methods of calculation selected.

2. Effect of Temperature, Pressure, and Composition of Gas

The effect of temperature will be reflected by the viscosity and density of the gas, and by the Cunningham factor, diffusivity, and Knudsen number of the particles.

For all gases the viscosity increases and the density decreases, as temperature increases. When the ratio $\nu_f = \mu_f/\rho_f$ appears, increasing temperature will increase this ratio. The mean free path λ of gas molecules also increases, so that Kn will increase as $\sqrt{T}$ for a fixed particle size.

The Knudsen number in turn fixed the value of C and Th. Because of the larger mean free path C will be greater at higher temperature. The net effect on C/μ will depend upon the relative changes in C and μ individually (see below).

Th is independent of Kn for Kn > 1 but is a complex function usually tending to decrease as Kn decreases below 1.0. Hence as temperature increases Th may tend to increase and level off for the coarser particles, but be unaffected for the finer particles.

All diffusional processes are more rapid at higher temperatures. Values of $\mathcal{D}$ for the particles, and D_{AB} for interdiffusion of two gases A and B become larger.

The effects of increasing pressure are as follows: the viscosity is not appreciably affected unless pressure is very high, then it will be increased. The density is increased in direct proportion to pressure. Consequently ν_f is decreased, at least up to moderate pressures. The mean free path is decreased, causing a decrease in C. Since Kn will decrease, Th will tend to decrease for coarser particles and remain constant for finer particles (Kn > 1). Particle diffusivity and gas diffusivity are decreased, at least for pressures up to 10-20 atm, but data are lacking or uncertain for the effect of still higher pressures.

All of the electrostatic properties (E_o, E, E_D) are essentially independent of temperature and pressure. Hence the effects on the several K parameters are primarily those of C/μ.

In the case of air, the only significant variation in composition is that of the water vapor content, or humidity. This is limited to a range between zero and saturation which depends upon temperature. The larger the humidity the smaller is the density and smaller the viscosity but the effects are not great. For precise calculations the correct property values corresponding to the humidity should be used.

In the case of a stack gas resulting from a combustion process the oxygen content of the air is replaced to a large extent by water vapor, carbon dioxide, and perhaps some carbon monoxide. Since the resulting mixture is still largely composed of nitrogen, and since the properties of the other gases tend to offset each other flue gas may be treated as though it were air for most calculations.

For other gases the properties must be obtained or estimated and used as they are. Some of the effects of temperature and pressure are illustrated in the sample calculations below.

Table 5 shows how all of these trends effect each of the parameters individually. The net effect of a change in conditions upon an an indivudual capture mechanism will depend upon the interrelationships of the parameters involved. For example, in the case of Brownian diffusion in viscous flow an increase in temperature will decrease Re and increase Pe^{-1}. The effects upon η_{BD} are in opposite directions ($Re_D < 1$), as may be seen from equations (39) or (40); the net effect will have to be found by calculation. For direct interception, the value of R is unaffected by temperature or pressure but, in viscous flow around a cylinder there will be an effect due to Re_D as shown in equation (21).

Where two or more mechanisms must be considered, the overall effect upon the total isolated target efficiency may be determined by calculation. Several examples of this have been published. Strauss and Thring [47] show the effect of increasing temperature upon II, DI,

Table 5

Effect of Temperature and Pressure Upon Parameters

Parameter	Increased T	Increased P (1)
Re_D	ν_f up, Re_D down (Effect appears only in laminar flow)	ν_f down, Re_D up
Stk	C up, μ_f up, Stk $\propto$ C/μ	C down, Stk down
R	No effect	No effect
Pe^{-1}	$\mathcal{D}$ up, Pe^{-1} up	down, Pe^{-1} down
N_{FT}	ν_f up, T^{-1} down Th up N_{FT} down	ν_f down, Th down, N_T down
N_{FD}	D_{AB} up, N_{FD} up	D_{AB} down, N_{FD} down
G·Stk	C up, μ up	C down, G·Stk down
K_{EM}, K_{EI}, K_{EC}, K_{ES}	C up, up all K's $\propto$ C/μ	C down, all K's down

Note: (1) Effect of P increase, say up to 10 atm. only.

and BD individually for a cylindrical target and particles ranging from 0.01 μm to 5 μm. Collection by both II and DI are reduced, while that for BD is increased. The net effect is not given, but if equation (56) is applied, it would appear that η_T decreases at higher temperatures for the larger particles, and increases for the smaller ones. Strauss and Lancaster [48] have shown the combined effect of pressure and temperature upon II for 1 μm particles on 10 μm fiber in carbon dioxide. Example (case IV) below illustrates how their results might have been obtained. These cases indicate that for II the effect of temperature on viscosity is greater than on the Cunningham factor, so that there is a net decrease in Stk at higher temperature.

Example 6. On the following pages are tabulated the results of sample calculations to illustrate the above procedures. For each case the set of conditions is stated at the top. Below is given the value of each dimensionless group and the corresponding isolated target efficiency. The overall efficiency is calculated by (56).

III. AERODYNAMIC CAPTURE IN ARRAYS OF TARGETS

Collectors of the target type (fiber filters, granular bed filters, scrubbers) always function with a large number of targets operating simultaneously. The behavior of an individual target in such an array is not likely to be the same as when the target is isolated.

To develop a mathematical model for the collector performance, we begin with the study of the isolated target as presented in Section II above. Next we consider how the presence of neighboring targets in the array may modify the behavior of a single (unisolated) target. Finally the overall performance of the entire array must be examined.

The definition of aerodynamic capture efficiency (by any mechanism) is modified to apply to each single target as it functions in the array. A total single, but not isolated, target capture efficiency due to all mechanisms is obtained. The overall efficiency of the collector consists of an appropriate summation of all the single target efficiencies.

A. Modified Fluid Velocity Patterns

The primary modification of isolated target behavior caused by the presence of neighboring targets is in the pattern of the fluid flow field around the targets. This has an effect not only upon the aerodynamic capture efficiency but also upon the pressure drop in the gas flowing through the array of targets. (This important matter will be dealt with in the chapters devoted to collector performance theory.)

To deal with the modified flow pattern, the so-called "cell theory" was developed and applied by Happel [49], and by Kuwabara [50] in the case of cylindrical targets, and by Happel [51] and Pfeffer [52] in the case of spherical targets.

The concept of the cell theory is that in a randomly spaced (but in the case of cylinders, parallel) array of targets of diameter D,

		Case I			Case II		
T	P	20°C	1 atm				
	μ_f	1.81×10^{-4}	poises			Same	
	ρ_f	1.205×10^{-3}	gm/cm^3				
	ρ_p	1.0	gm/cm^3			as	
	v_o	15 cm/s.					
Shape	D	Cylinder	50 μm			Case I	
	d_p	10 μm			0.1 μm		
	C	1.02	Fig.3.3		2.67		Fig. 3.3
		2.3×10^{-8}	cm^2/s	(3.58)	6.7×10^{-6}	cm^2/s	(3.58)
	q	300 e			300 e		
	ε_D	10			10		
	Re_D	0.50	η		0.5	η	
II	Stk	0.95	0.55	(29a)	2.4×10^{-4}	-	
DI	R	0.20	0.013	(32)	0.002	-	
BM	Pe^{-1}	3.2×10^{-7}	6×10^{-5}	(39)	8.9×10^{-5}	2.5×10^{-3}	(39)
G	G•Stk	0.02	0.020	(54)	5.3×10^{-6}	-	
EM	K_{EM}	3.4×10^{-4}	0.022	Tab.3	0.089	0.36	Tab. 3
			$\eta_T = 0.57$	(56)		$\eta_T = 0.36$	
		II predominates			Electrostatic predominates		

		Case III			Case IV			
T	P	300°C	1 atm		300°C		10 Atm	
	μ_f	2.93×10^{-4}	poises			2.93×10^{-4}	poises	
	ρ_f	0.613×10^{-3}	gm/cm^3			0.613×10^{-2}	gm/cm^3	
	ρ_p	1.0	gm/cm^3			2.5	gm/cm^3	
	v_o	15 cm/s.				1 cm/s		
Shape	D	Cylinder 50 μm			Sphere	50 μm		
	d_p	0.1 μm			10 μm			
	C	5.56	Fig. 3.3		1.003	(3.6)		
		$1.7 \times k0^{-5}$	cm^2/s	(3.58)	1.85×10^{-8}	(3.58)		
	q	0			0			
	D	-			-			
	Re_D	0.16	η		0.105	η		
II	Stk	3.2×10^{-4}	-		0.095	0.033	(29b)	
DI	R	0.002	-		0.20	0.057	(33)	
BM	Pe^{-1}	2.3×10^{-4}	0.00050	(41)	3.7×10^{-6}	7×10^{-4}	(43)	(44)
G	G Stk	6.9×10^{-6}	-		0.47	0.47	(54)	
EM	K_{EM}	0	-	-	-	-		
		$\eta_T = 0.005$			$\eta_T = 0.49$			
		BM predominates; increases with temperature			Gravity predominates at high ρ_p and low v_o			

each target is surrounded by an imaginary concentric surface of similar shape and radius b, which encloses fluid and comprises a bounding free surface or cell. The value of b is determined by the porosity ε (fractional void volume) of the array such that $\varepsilon = 1-D^2/4b^2$ (cylinders), or $\varepsilon = 1-D^3/8b^3$ (spheres), and $c = 1-\varepsilon$.

Happel assumed that there was zero shearing stress at the cell boundary, Kuwabara that there was zero vorticity. Both solved the Navier-Stokes equation for viscous flow in the fluid between the target boundary and the cell boundary, assuming there was no slip on the target surface. Similar results were obtained in terms of the effect of porosity to be incorporated into the velocity component equations for viscous flow around a single target.

For flow tranverse to cylinders, Kuwabara obtained

$$v_r = -\frac{v_o}{2Ku}\left[2\ln\frac{2r}{D} - 1 + c + \frac{D^2}{4r^2}\left(1 - \frac{c}{2}\right) - \frac{2cr^2}{D^2}\right]\cos\theta \qquad (59)$$

$$v_\theta = \frac{v_o}{2Ku}\left[2\ln\frac{2r}{D} + 1 + c - \frac{D^2}{4r^2}\left(1 - \frac{c}{2}\right) - \frac{6cr^2}{D^2}\right]\sin\theta$$

in which

$$K_u = -\frac{1}{2}\ln c - \frac{3}{4} + c - \frac{c^2}{4} \qquad (60)$$

and $v_o = V/\varepsilon$, where V is the superficial velocity of gas approaching the array.

Kuwabara's results for cylinders seem to agree best with experimental tests according to Davies [53, pg. 54]. As $\varepsilon \to 0$ ($c \to 1$) these equations reduce to those of Lamb (26) for an isolated target provided Ku is replaced by ($2.002 - \ln Re_D$). At a value of $\varepsilon = 0.001$ the Kuwabara field is identical to the Lamb field for $Re_D = 0.495$, when Ku = 2.705. It is remarkable that in the Kuwabara viscous field there is no effect of Re_D, nor of fluid viscosity, only of the porosity of the array. Crawford [28, pg. 432] has provided a chart for Ku as a function of ε.

For spherical targets the Happel free surface model [51] has been applied by Pfeffer [52] to obtain

$$v_r = \frac{v_o \cos\theta}{W}\left[2 + 3\gamma^5 \pm \frac{a^3}{r^3} - (3 + 2\gamma^5)\frac{a}{r} - \frac{\gamma^5 r^2}{a^2}\right]$$
$$v_\theta = \frac{v_o \sin\theta}{2W}\left[4 + 6\gamma^5 - \frac{a^3}{r^3} - (3 + 2\gamma^5)\frac{a}{r} - \frac{4\gamma^5 r^2}{a^2}\right] \qquad (61)$$

in which $\gamma^3 = 1 - \varepsilon = c$ and $W = 2 - 3\gamma + 3\gamma^5 - 2\gamma^6$.

B. Single Target Behavior in Arrays

In principle, single target capture efficiency should be calculated for arrays in the same manner as for isolated targets except for using the modified fluid velocity pattern. For example, equations (59) would be substituted for v_r and v_θ in equations (24) in order to determine inertial trajectories. In practice, the matter becomes very complex.

For arrays of cylinders Davies [53] has devoted an entire chapter to the application of the Kuwabara field to aerodynamic capture efficiency calculations. He recommends, finally, the following scheme for evaluating the overall single target efficiency incorporating inertial impaction, direct interception, and Brownian diffusion.

$$\eta_T \equiv \eta_{DI} + \eta_{II} + \eta_{BD} \qquad (62)$$

where

$$\eta_{DI} = \frac{1}{2Ku}\left[2\,(1 + R)\ln(1 + R) - (1 + R) + (1 + R)^{-1} + c\left(-2R^2 - \frac{R^4}{2} + \frac{R^5}{2} + \ldots\right)\right] \qquad (63)$$

$$\eta_{II} = \frac{J}{Ku}\,\mathrm{Stk} \qquad \text{(see Table 6)} \qquad (64)$$

$$\eta_{BD} = \frac{2.9}{Ku^{1/3}\ Pe^{2/3}} + \frac{0.62}{Pe} + \frac{1.24\ R^{2/3}}{Ku^{1/2}\ Pe^{1/2}} \tag{65}$$

The last term in η_{BD} represents a combination effect of diffusion and interception. The J/Ku values in η_{II} depend only upon c and R and may be obtained from Table 6 which is recalculated from Davies.

The limitations on this set of equations, given by Davies [53], are as follows:

$v_o = V/\varepsilon = V/(1-c)$ (velocity inside array)

$Re_D = \dfrac{v_o D}{\nu_f} < 0.5$ (viscous flow thru array)

$R < 0.1$

$Pe > 100$

$Stk < 0.1$ (C must be included)

Note that the calculations of the individual η's have been defined in such a way that they are merely added to get η_T and not combined as suggested by (56).

Example 7. Estimate the total single target efficiency under isothermal conditions for dry air at 20°C, 1 atm, flowing at v = 10 cm/sec through an array of cylinders each 10 microns in diameter packed with a porosity of 95% upon spherical particles of unit density and 0.01, 0.1, 1, and 2 microns in diameter respectively. Compare the result with that for an isolated cylinder under the same conditions.

For the array, c = 0.05; Ku = 0.797 (equation 60), $Re_D = 0.0664 < 5$.

Parameter values:

d_p − μm	C	Stk	R	J/Ku	$\mathcal{D}$ $\frac{cm^2}{sec}$	Pe
0.01			0.001	-	5.3×10^{-4}	18.9
0.1	2.89	0.000887	0.01	0.000314	6.8×10^{-6}	1470
1.0	1.166	0.0357	0.1	0.268	2.76×10^{-7}	3.6×10^{4}
2.0	1.083	0.1326	0.2	0.880	1.3×10^{-7}	7.7×10^{4}

Efficiencies:

d_p - m	η_{DI}	η_{II}	η_{BD}	η_T
0.01	-	-	0.476	0.476
0.1	0.0002	-	0.0324	0.0326
1.0	0.0105	0.0096	0.0045	0.0245
2.0	0.419	0.117	-	0.159

Note: η_{DI} from (63), η_{II} from (64), η_{BD} from (65), η_T from (62).

Table 6
Values of J/Ku for Equation (64)

c	R = 0.01	0.02	0.05	0.1	0.2
0.005	0.00152	0.00588	0.0352	0.1276	0.434
0.01	0.00180	0.00678	0.0416	0.1510	0.514
0.02	0.00220	0.00816	0.0510	0.1852	0.626
0.05	0.00314	0.01180	0.0730	0.268	0.880
0.1 0.	0.00460	0.01704	0.1056	0.386	1.248*

* Impossible value

For an isolated cylinder, the parameters are the same except that c = 0; Ku, and J/Ku do not apply. Equation (57) as modified, may be used for η_T, although it is for $Re_D = 0.2$

Efficiencies:

d_p - μm	η_T	η_{BD}	η_{II}	η_{DI}	η_T
0.01	0.0044	0.1438	-	-	0.144
0.1		0.0079	-	0.0002	0.008
1.0	0.0045	0.0009	0.01	0.0020	0.013
2.0	0.046	-	0.02	0.0075	0.027

Note: η_{DI} form (32), η_{II} from Fig. 7, η_{BD} from (39) and

$$\eta_T = 0.16\ [R + (0.50 + 0.8R)\ (Stk + 1/Pe) - 0.1052R\ (Stk + 1/Pe)^2]$$

To determine the effect of electrostatic charge upon arrays of cylinders, Lundgren and Whitby [54] performed experiments using charged particles and uncharged fibrous filters. They determined the single fiber efficiency caused by the charging itself using equation (56) and correlated it with the corresponding parameter K_{EM} developed by Natanson [37] and given in Table 3. The parameter given in the table is 4π times the K_M quoted by Lundgren and Whitby. Their results for viscous flow fit well with:

$$\eta_E = 1.5\ K_M^{1/2} \tag{66}$$

where $K_M = K_{EM}/4\pi$ and $(\varepsilon_D - 1)/(\varepsilon_D + 1$ was taken as 1. The value of η_E would have to be incorporated with other single target efficiencies using equation (56) in order to get the total efficiency when charged particles are present.

For spherical targets Paretsky, Theodore, Pfeffer and Squires [55] have used Pfeffer's concepts and equations (61) to calculate single target efficiencies as follows:

$$\eta_{DI} = 3R^2\ \rho(\varepsilon)^{-1} \tag{67}$$

$$\eta_{BD} = 5.04\ Pe^{-2/3}\ \rho(\varepsilon)^{-1/3} \tag{68}$$

$$\eta_G = 0.0624\ \frac{u_s}{v_o} = 0.125\ G \cdot Stk \tag{69}$$

in which

$$\rho(\varepsilon) = \frac{2 - 3(1-\varepsilon)^{1/3} + 3(1-\varepsilon)^{5/3} - 2(1-\varepsilon)^2}{1 - (1-\varepsilon)^{5/3}} \tag{70}$$

Note that $\rho(\varepsilon) \to 2$ as $\varepsilon \to 1$, and that $\eta_{DI} \to 3R^2/2$, equation (32), and $\eta_{BD} \to 4.0\ Pe^{-2/3}$, equation (45). $\rho(\varepsilon)$ appears to be valid only for values of ε near 1 because it goes through a minimum near $\varepsilon = 0.78$.

η_{II} is given by a graph, similar to Fig. 8, plotted as a function of (2Stk) with curves for ε = 0.39, 0.43, 0.49, 0.66, 1, the last of which coincides with Hern's [12] calculation for isolated spheres. At a given value of Stk, the lower the porosity, the higher is η_{II}. These curves all predict a limiting low value of 2Stk ≃ 0.1 below which η_{II} is zero. However some experimental data show collection to occur in the range of $10^{-3} < \eta_{II} < 10^{-2}$ for Stk ≃ (1-4) x 10^{-3} at $\varepsilon \simeq 0.42$.

C. Overall Behavior of Array

Finally the overall efficiency of an entire array of targets must be related to the single target efficiency and the number of targets present. An elementary way to do this employs the lateral-mixing model outlined in Chapter 3.

Consider an array of N_T targets per unit volume, each target having projected area of A_T in the direction of flow. In a unit cross-sectional area of the array, having a thickness of dL in the direction of flow, there will be a total of $N_T \cdot dL$ targets and $A_T N_T \cdot dL$ total projected area. Assuming all the targets to be alike and to be arranged in a regular fashion, the total single target capture efficiency η_T may be applied to every one in parallel. Then the number of particles approaching the target per unit time will be nv, and the reduction will be $v_o dn$, which must be

$$-v_o dn = nv \times \eta_T \times A_T N_T dL \tag{71}$$

Assuming η_T. A_T, N_T, and ε to be constant, integration gives the total capture efficiency of thickness L of the array

$$-\int_{n_o}^{n} \frac{dn}{n} = \frac{\eta_T A_T N_T}{\varepsilon} \int_o^L dL \tag{72}$$

$$-\ln P = \eta_T A_T N_T L/\varepsilon \qquad \text{or} \qquad \eta = 1 - \exp - \eta_T A_T N_T L/\varepsilon \tag{73}$$

The dimensionless quantity $A_T N_T L/\varepsilon$ is sometimes called the solidarity factor S of the array.

For an array of cylindrical targets, each D by ℓ, packed with porosity ε

$$A_T = D\ell$$

$$N_T = (1-\varepsilon)/\frac{\pi D^2 \ell}{4}$$

Then

$$\eta = 1 - \exp -\eta_T\ (1-\varepsilon)\ 4L/\varepsilon\pi D \text{ and } S = 4(1-\varepsilon)L/\pi\varepsilon D \tag{74}$$

When this is applied to an array of fibres, such as in a fibrous filter pad, D may be taken as the average fibre diameter and a fictitious length ℓ' may be defined to represent the total "effective" fibre length in a unit cross-section of filter. Thus $L = \pi D^2 \ell'/4(1-\varepsilon)$ and by substitution

$$\eta = 1\text{-}\exp\text{-}\eta_T D\ell'/\varepsilon \tag{75}$$

Since the fibres in such a pad are not in a regular array, i.e., not all parallel to each other and perpendicular to the flow, ℓ' may be taken to represent the equivalent effective total length they would have if they were so arrayed.

For an array of spherical targets, each of diameter D:

$$A_T = \pi D^2/4$$

$$N_T = 6(1-\varepsilon)/\pi D^3$$

Then

$$\eta = 1\text{-}\exp - 3(1-\varepsilon)\eta_T L/2\varepsilon D \quad \text{and} \quad S = 3(1-\varepsilon)L/2\varepsilon D \tag{76}$$

In case the elements of the array are spaced far from each other, as drops in a spray scrubber might be, there is no interaction effect.

But the stream of particles may be subjected to possible capture by a number η of targets in series, each acting as though isolated. In that case the overall efficiency will simply be a repeated application of (56), in which η_T is an isolated target efficiency.

$$\eta = 1-(1-\eta_T)^n \tag{77}$$

or

$$P = P_T^{\ n}$$

If n is sufficiently large and η_T relatively small this will approach

$$\eta = 1-\exp{-n\eta_T} \tag{78}$$

as may be shown by comparing the series expansions for (77) and (78) term by term.

In all cases the function of the array is to multiply the capture effect of a single target. Even though the single target efficiency be rather low, the overall efficiency of an array may be substantial.

Example 8. Using the results of Example 7 calculate the grade efficiency and number of transfer units of an array of cylinders each 10 microns in diameter, packed with a porosity of 95%, and either 1 mm or 2 mm in thickness

$$S = \frac{4\ (0.05)}{0.95}\ \frac{L}{10\text{x}10^{-4}} = 67.0\ L$$

$$\eta_i = 1-\exp - \eta_{T_i}\ (67.0)L$$

$$NTU_i = -\ln\ (1-\eta_i) = 67.0\ L\eta_{T_i}$$

where

$$L = 0.1\text{ cm, or } 0.2\text{ cm}$$

Grade Efficiency η_i

d_p - μm	η_{T_i}	L=0.1 cm	L=0.2 cm	(NTU) i L=0.1	(NTU) i L=0.2
0.01	0.476	0.959	0.9983	3.19	6.38
0.1	0.0326	0.196	0.354	0.218	0.436
1.0	0.0245	0.151	0.280	0.164	0.328
2.0	0.159	0.655	0.881	1.06	2.12

Note how the dominating collection mechanism shifts from Brownian diffusion to inertial impaction as the particle size increases, with an intermediate range in which direct interception is predominant although no mechanism produces much collection. The number of transfer units is directly proportional to the thickness of the array.

Calculations such as these are only valid for the initial period of collection by a stationary array. As collected particles accumulate on the target elements, the shape and nature of the targets changes. Particles collect upon each other and may build up layers or "dendrites" extending out from the original targets. Subsequent collection is a more complex phenomenon and has not been analyzed completely.

REFERENCES

1. N.A. Fuchs, *The Mechanics of Aerosols,* Pergamon Press, New York 1964.

2. J.D. Tuttle, M.S. Thesis, University of Cincinnati, Cincinnati, Ohio 1976.

3. H. Lamb, *Hydrodynamics,* 6th ed., Dover, New York, 1945.

4. W. Strauss, *Industrial Gas Cleaning,* 2nd ed. Pergamon Press, New York, 1975.

5. W.E. Ranz, *Bull. No. 65,* Dept. of Eng. Res., The Pennsylvaina State Univ., University Park, Penn. 1956.

6. M.N. Golovin and A.A. Putnam, *Ind. Eng. Chem. Fund.,* 1: 264 (1962).

7. K.R. May and R. Clifford, *Ann. of Ind. Hyg,* 10: 83 (1967).

8. R.J. Brem and H.W. Mergler, *NACA* TN 2904, March 1953.

9. C.N. Davies and M. Aylward, *Proc. Phys. Socl (London)*, B64: 889 (1951).

10. H.D. Landahl and R.G. Hermann, *J. Colloid Sci.*, 4: 103 (1949).

11. R.G. Dorsch, P.G. Sapes, C.F. Kadow, *NACA* TN 3099, March 1954.

12. H. Herne, *Int. J. of Air Poll.*, 3: 26, (1960).

13. H.F. George and G.W. Poehlein, *Environ. Sci. & Tech*, 8: 46 (1974).

14. T.T. Mercer and R.G. Stafford, *Ann. Occup. Hyg*, 12: 41 (1969).

15. S.C. Stern, *Ind. Eng. Chem. Fund.*, 1: 273 (1962).

16. T.T. Mercer and H.Y. Chow, *J. Colloid & Interrac. Sci.*, 27: 75 (1963).

17. M.K. Householder and V.W. Goldschmidt, *J. Colloid and Interfac. Sci.*, 31: 464 (1969).

18. Y. Cheng and C.S. Wang, *J. Aerosol Sci.*, 6: 139 (1975).

19. R.I. Crane and R.L. Evans, *J. Aerosol Sci.*, 8: 161 (1977).

20. S. Gadzioch, *J. Inst. Fuel*, 33: 106 (1960).

21. J.I.T. Stenhouse, P.J. Lloyd, *A.I.Ch.E. Symp. Ser., no. 137*, 70: 307 (1974).

22. K.T. Whitby, *ASHRAE J.*, 7: 56 (1965).

23. S. Calvert, J. Goldschmid and D. Leith EPA - 650/2-73-936, (1973).

24. J. Pich in *Aerosol Science* (C.N. Davies, Ed.), Academic Press New York, 1966, Chap. 9.

25. I. Langmuir, ORSD Report 865, 1942.

26. G.L. Natanson, *Dokl. Akod. Nauk SSSR*, 112: 100 (1957).

27. S.K. Friedlander, *A.I.Ch.E.J.*, 3: 11 (1957).

28. M. Crawford, *Air Pollution Control Theory*, McGraw-Hill New York, 1976.

29. W.L. Torgenson in *General Mills* Rep. No. 1890 AEC Cont. No. AT-(11-1) -401, SC. Stern et al (1958).

30. W.E. Ranz, Tech. Rep. No. 3, Contract AT - (30-3)-28, University of Illinois (1951).

31. C.J. Stairmand, *Trans. Inst. Chem. Engrs. (London),* 28: 130 (1950).

32. H.F. Johnstone and M.H. Roberts, *Ind. Eng. Chem.,* 41: 2417 (1949).

33. R. Pfeffer, *Ind. Eng. Chem. Fund.,* 3: 380 (1964).

34. M.J. Pilat and A. Prem, *Atmos. Environ.,* 10: 13 (1976).

35. L.E. Sparks and M.J. Pilat, *Atmos. Environ.,* 4: 651 (1970).

36. H.F. Kraemer and H.F. Johnstone, *Ind. Eng. Chem.,* 47: 2426 (1955).

37. G.L. Natanson, *Dokl. Akad. Nauk SSSR,* 112: 696 (1957).

38. W.E. Ranz and J.B. Wong, *Ind. Eng. Chem.,* 44: 1371 (1952).

39. J.W. Thomas and R.E. Yoder, *A.M.A. Arch. Ind. Health,* 13: 545 (1956).

40. C.N. Davies, *Proc. Inst. Mech. Engrs.,* 1B: 185 (1952).

41. S.K. Friedlander and R.E. Pasceri, *Can. J. Chem. Eng.,* 38: 212 (1960).

42. C.E. Billings, *Effects of Particle Accumulation in Aerosol Filtration,* California Inst. of Technology, Pasadena, Sept. (1966).

43. H.F. George and G.W. Poehlein, *Environ. Sci. and Tech. vol,* 8: 46 (1974).

44. I. Langmuir and K. Blodgett, *Army Air Forces Tech. Report:* No. 5418, (1945).

45. R.T. Jarman, *J. Agric. Eng'g. Res.,* 4: 139 (1959).

46. H. Emi and N. Yoshioka, Prediction of Collection Efficiencies of Aerosols by Fibrous Filters, *First Pacific Chem. Eng. Congress,* Oct. 10-14, 1972, Kyoto, Japan.

47. W. Strauss and M.W. Thring, *Trans. Inst. Chem. Engrs.* 41: 28 (1963).

48. W. Strauss and B.W. Lancaster, *Atmos. Environ.* 2: 135 (1968).

49. J. Happel, *Am. Inst. Chem. Eng. J.,* 5: 174 (1959).

50. S. Kuwabara, *J. Phys. Soc. Japan,* 14: 527 (1959).

51. J. Happel, *Am. Inst. Chem. Eng. J.,* 4: 197 (1958).

52. R. Pfeffer, *Ind. Eng. Chem. Fund.,* 3: 380 (1964).

53. C.N. Davies, *Air Filtration,* Academic Press, London (1973).

54. D. Lundgren and K.T. Whitby, *Ind. Eng. Chem. Proc. Des. Dev.,* 4: 345 (1965).

55. L. Paretsky, L. Theodore, R. Pfeffer, and A.M. Squires, *J. Air Poll. Cont. Assoc.* 21: 204 (1971).

56. P. Knettig and J.M. Beeckmans, *Aerosol Sci.* 5: 235 (1974).

57. K.A. Nielsen and J.C. Hill, *Ind. Eng. Chem. Fundam.,* 15: 149 (1976).

58. K.A. Nielsen and J.C. Hill, *Ind. Eng. Chem. Fundam.,* 15: 157 (1976).

59. S. Calvert and R. Parker PB 264203, EPA 600/7-77-202 (1977).

60. K. Willeke, *Jour. Aerosol Sci.,* 7: 381 (1976).

CHAPTER 6

CENTRIFUGAL COLLECTORS

I. CYCLONE DESIGN AND OPERATION

Collectors in which the major collecting force is centrifugal in nature are of a type known as cyclones, although this kind of force may be a factor in some other types as well. The calculations to be discussed here are primarily applicable to a vertical reverse flow cyclone with a tangential inlet. This is shown schematically in Figure 1. Variations on this design involve a scroll inlet, or a helical inlet instead of the tangential duct. Straight-through flow cyclones, without the conical portion, are also used, usually in a horizontal position. Recent innovations in design include skimming off a portion of the outlet gas for recycling, and the use of a secondary air stream. These variations are discussed at the end of this chapter.

Referring to Figure 1, the dust-laden gas enters tangetially through the duct, shown as rectangular with height a and width b. This imparts the spinning motion which directs the dust particles radially outward toward the collecting surface, primarily the cylinder of diameter D and height h. As the gas stream spins below the level of the outlet (exit) duct, which is a cylinder of diameter D_e

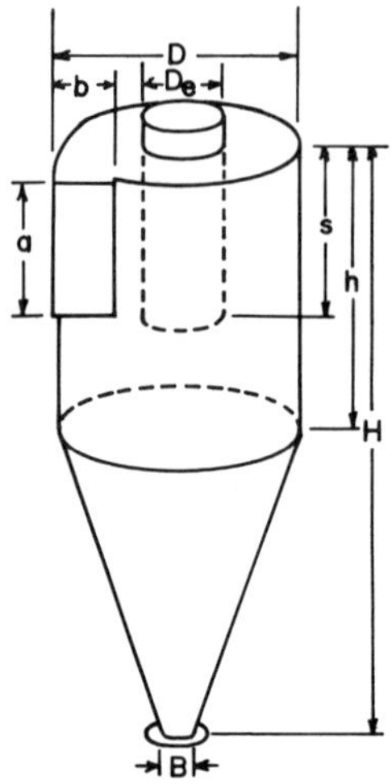

FIG. 1. Cyclone shape and dimensions.

and height S (sometimes referred to as the "vortex finder" or the "tubular guard"), a radially inward motion of gas develops. Swirling motion of the gas in a helical pattern continues into the conical base, of height (H-h), but eventually all of the gas flows inward, upward and out through the outlet duct. This gives rise to the name "reverse-flow" for this type of design. Meanwhile the dust colliding with the outer walls in the upper cylinder, as well as to some extent in the lower conical portion, slides down these walls (aided by a powerful Taylor boundary layer secondary flow) and out through point B into a collecting hopper below.

The collecting force developed is such as to be particularly effective on larger particles, say upwards of 10 microns in equivalent diameter. The collection velocity is counteracted to some extent by bouncing of coarser particles off the walls, and by reentrainment of particles while they are sliding downward. At high tangential velocity the "saltation" effect may inhibit collection. Conditions are very turbulent and there is considerable back-mixing of particles radially inward. A typical experimental grade-efficiency curve [1] may be seen in Fig. 2.

There appears to be no practical upper limit to dust loading which a cyclone may handle. Loadings of over 1000 gr/ft^3 have been treated. Higher grain loading seems to bring about simultaneously an increase in collection efficiency and a decrease in pressure drop.

The device is relatively simple in design, construction and operation, having no mechanical parts except the fan required to move the gas stream. It is a relatively inexpensive collector, both in initial cost and operating cost. By using suitable materials and methods of construction it may be adapted to extreme operating conditions: high temperature, high pressure, corrosive gases. It is not, however, suitable if a high degree of collection of very fine particles is needed. It may well serve as a pre-cleaner to remove the coarser particles from a gas which is subsequently to be cleaned of fine particles by one of the other devices. Two or more cyclones may also be operated in a series or in parallel upon a given gas-stream.

To select or design a cyclone for a specific duty requires specification of its overall size, characterized by the principal diameter D, and its shape or configuration which is determined by seven geometric ratios:

$$\frac{a}{D}, \frac{b}{D}, \frac{D_e}{D}, \frac{S}{D}, \frac{h}{D}, \frac{H}{D} \text{ and } \frac{B}{D}$$

Selection of these eight quantities will in turn depend upon the duty to be performed, as described by:

- Rate of flow, temperature, pressure of gas stream
- Composition of the gas, including especially its humidity
- Dust loading of gas
- Size-distribution of dust carried by gas
- Degree of total collection, or reduced level of emission, to be attained

Details of the design, and selection of material of construction may also be influenced by the chemical and physical nature of the dust.

In order to develop an optimal design, it is necessary to have a model for the collection process which gives the grade-efficiency function, and a method of calculating the pressure drop through the device. Ideally, these models must show how the eight geometrical parameters and all the operating variables will affect the grade-efficiency and the pressure drop.

II. MODELING CYCLONE PERFORMANCE

A. Radial Mixing

Earlier attempts to model the collection process [2, 3, 4], were based essentially upon some scheme similar to the no-mixing model described in Chapter 3. These were not very successful, as they tend to produce S-shaped grade-efficiency curves such as are illustrated in Fig. 2.

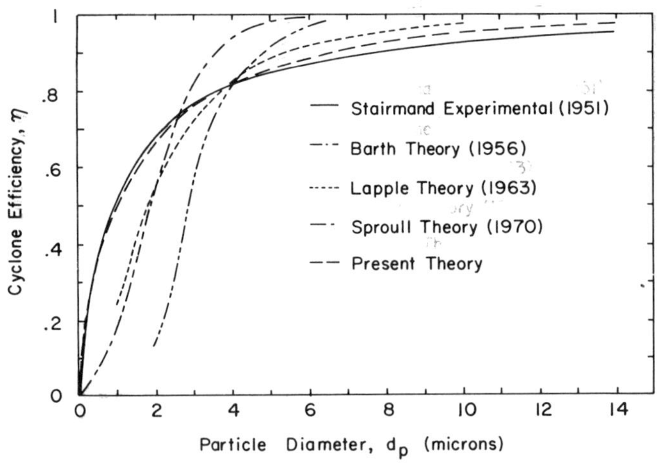

FIG. 2. Experimental grade efficiency [1] and model tests [6]. Reproduced by permission from American Institute of Chemical Engineers, Symposium Series.

The model to be described and used here, is based upon the radial-mixing concept, similar to the lateral-mixing model presented in Chapter 3. It has given the best agreement with the available experimental data which is sufficiently complete for checking [6].

The model is predicated upon the observation that, in spite of the centrifugal force acting upon the particles, a radially-inward mixing of them will tend to occur because of the turbulence and eddies in the vortex of gas, because of drag on the particles by the inward flowing gas below the level of S, and because of particle-bouncing off the cylinder walls. It is postulated that these effects are sufficient to maintain a uniform concentration of uncollected dust in any horizontal cross-section of the cyclone, i.e., that radial-mixing is complete.

Consider a horizontal cross-section, as shown, just below the level S of the outlet duct:

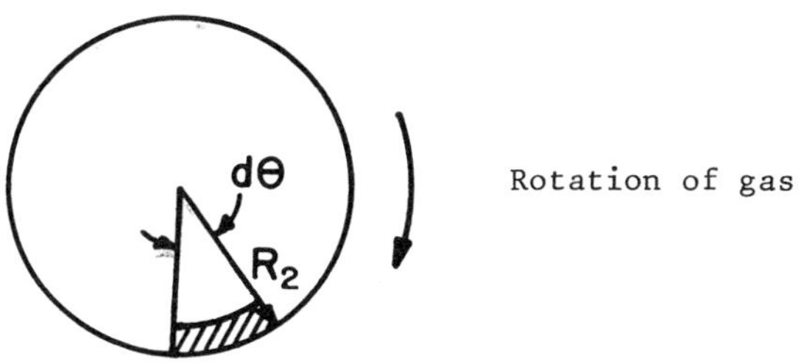

In a period of time dt, only those particles within a certain distance dR of the cyclone wall (indicated by radius R_2) will move to the wall and be collected. Meanwhile the particles will travel a distance $R_2 d\theta$ tangentially and dL vertically. The number of particles removed (dn') will be

$$-dn' = d\theta/2 \left[R_2{}^2 - (R_2 - dR)^2\right] n\ dL \qquad (1)$$

The total number n' of particles in the sector from which these particles are removed is

$$n' = d\theta/2\ {R_2}^2\ n\ dL \tag{2}$$

The fraction of particles removed in time dt is therefore

$$-\frac{dn'}{n'} = \frac{2R_2 dR - (dR)^2}{{R_2}^2} \simeq \frac{2dR}{R_2} \tag{3}$$

On the right-hand side of (3), the second-order differential is neglected. It is true, of course, that in the upper part of the cyclone, between the levels S and a, the particle system cannot be physically distributed all the way into the center because of the presence of the outlet duct. Since (S-a) is usually small however, this correction has been neglected.

In order to relate the fraction of particles collected to the total average residence time, it is necessary to express (3) in terms of time. This may be done through equation (18), Chapter 5, which was developed to show the relationship between time and the radial position of a single particle in a vortex. A system of uncollected particles, evenly distributed across the cross-section of such a vortex, will have its center of mass at the vortex center. If the system of uncollected particles is instantaneously, continuously, and completely redistributed by radial-mixing as postulated, the center of mass of the uncollected particle system will remain at the vortex center, even though the total number of uncollected particles is decreasing with increasing gas residence time.

The rate at which the uncollected particle system moves toward the vortex wall, as a function of the time the system has spent within the vortex, may then be obtained by saying that the radial position of the uncollected particle system is at the vortex center, i.e., set $R_1 = 0$ in (18), Chapter 5, solve for R and differentiate with respect to t, to obtain

$$\frac{dR}{dt} = \frac{\tau u_{T_2}^2}{R_2}\left[2(n+1)\ \tau\left(\frac{u_{T_2}}{R_2}\right)^2 t\right]^{-\frac{2n+1}{2n+2}} \tag{4}$$

Combining (4) with (3) and integrating over the residence time t:

$$\int_{n_o'}^{n'} \frac{dn'}{n'} = -\frac{2\tau u_{T_2}^2}{R_2^2}\left[2(n+1)\,\tau\left(\frac{u_{T_2}}{R_2}\right)^2 t\right]^{-\frac{2n+1}{2n+2}} \int_0^t t^{\left(-\frac{2n+1}{2n+2}\right)} dt \quad (5)$$

$$\eta_i = 1 = \exp - 2\left[2(n+1)\,\tau_i\left(\frac{u_{T_2}}{R_2}\right)^2 t\right]^{\frac{1}{2n+2}} \quad (6)$$

This gives the grade-efficiency of collection as a function of the residence time.

In order to use (6), the calculation of each of the terms in it must be specified. Since R_2 has already been taken as the position of the outer wall, $R_2 = 1/2\ D$. The value of u_{T_2} should correspond to the same location, but strictly speaking $u_{T_2} = 0$ at R_2 because there is a boundary layer of collected dust on the wall. It is customary to replace u_{T_2} by the average inlet velocity u_T through the entrance duct, $u_T = Q/ab$.

The value of n may be estimated by a method suggested by Alexander [7] which is an empirical correlation of his own experimental measurements:

$$n_i = 0.67\ (D)^{0.14} \quad \text{at } 10°\text{C (D in m)}$$

and

$$(1-n_T)/(1-n_i) = (T/283)^{0.3} \quad \text{at } T°\text{K}$$

combining:

$$n_T = 1 - \left[1 - 0.67\ (D)^{0.14}\right](T/283)^{0.3} \quad (7)$$

D must be expressed in meters.

B. Residence Time

There remains the question of the proper value of t to use. This is determined by the pattern of gas flow through the cyclone. The gas enters over a span of height ranging from the top of the cyclone vertically downward to the depth a. As it passes vertically downward below the level S, it is gradually drawn off into the central core along the axis of the exit duct. Ter Linden's measurements [9] show that this flow is at a lineal velocity which is relatively constant, regardless of radial position or elevation. Different parcels of gas introduced into the cyclone at the same time may thus have different residence times within the unit, depending upon the level at which they are introduced and upon how soon after entry they are drawn off. The range of residence times would be from the shortest, corresponding to entry at level a and exit at level S, to the longest which would correspond to entry at the top and exit at depth H.

For simplicity it is desirable to determine an average residence time for all of the gas stream such as will account for the collection obtained. It has been found that an adequate way to do this is as follows. The total average residence time may be taken to equal the average time required for gas to descend from the average level of entrance to the level of the bottom of the exit pipe, plus the average time in residence below this point. The average minimum residence time can be calculated by assuming all the gas enters the cyclone at the midpoint of the entrance duct. With this assumption,

$$t_{min_{avg}} = \frac{\pi(S-a/2)\ (D^2 - D_e{}^2)}{4Q} = \frac{V_s}{Q} \tag{8}$$

where

$$V_s = \pi(S-a/2)\ (D^2 - D_e{}^2)/4 \tag{9}$$

The additional time of residence will vary from zero to a maximum corresponding to the lowest point of descent of the gas into the conical cyclone body, with the average additional time being taken as one-half of this maximum.

The lowest point of descent of gas is not necessarily the actual bottom of the cyclone as measured by the depth dimension H. Alexander [7] has observed with very long glass cyclones that the gas vortex will have a "well-defined stable turning point" at some distance below the bottom of the exit duct which is less than (H-S). He calls this the "natural length" ℓ of the cyclone and gives an empirical formula for it, as:

$$\ell = 2.3\ D_e (D^2/ab)^{1/3} \tag{10}$$

It is noteworthy that this natural length is, according to Alexander, independent of the gas flow rate.

The maximum additional residence time, of gas not drawn off until reaching the tip of the vortex, will be:

$$t_{max} = V_{n\ell}/Q \tag{11}$$

where $V_{n\ell}$ is the effective volume of the lower portion of the cyclone at the natural length, where the vortex turns. This effective volume is equal to the total volume of the cyclone from the level of the exit duct down to the level of the natural length, minus the volume of the "central core" of the cyclone where the gas stream is swept upwards and out of the particle-separating vortex. Taking the diameter of the central core to be equal to that of the gas exit duct, gives

$$V_{n\ell} \quad \frac{\pi D^2}{4} (h-S) + \frac{\pi}{4} \frac{(\ell + S-h)\ D^2}{3} \left[1 + \frac{d}{D} + \frac{d^2}{D^2} \right] - \frac{\pi D_e^{\,2} \ell}{4} \tag{12}$$

where

$$d = D - (D-B)\ (S+\ell-h)/(H-h)$$

However, the core diameter, as a fraction of the exit pipe diameter has been variously given as 1/2, 2/3 and 1 by Stairmand [1], Ter Linden [9], and Barth [2] respectively.

A practical cyclone should have a physical length (H-S) near its natural length ℓ. If the cyclone body is longer than ℓ, i.e., $(H-S) > \ell$, the space at the bottom of the cyclone, below the vortex turning point, will be wasted. If the cyclone body is shorter than ℓ, i.e., $(H-S) < \ell$, the full separating potential of the cyclone will not be realized. In this case t_{max} will be determined simply by the total lower volume of the cyclone minus that of the "central core." That is, $V_{n\ell}$ would be replaced by V_H which is:

$$V_H = \frac{\pi D^2}{4}(h-S) + \frac{\pi D^2}{4}\frac{(H-h)}{3}\left[1 + \frac{B}{D} + \frac{B^2}{D^2}\right] - \frac{\pi D_e^{\,2}}{4}(H-S) \tag{13}$$

Aslami [10] has found that it is possible to have a design in which $D_e > d$, or in which $D_e > B$, such that $V_{n\ell}$ or V_H (respectively) will be negative. In this case he recommends taking the central core as the frustum of a cone extending from D_e at S to d at ℓ, or B at H, as the case may be.

The average total residence time of the gas in a cyclone will then be taken as

$$t_{res} = (V_s + V_{n\ell}/2)/Q = K_c D^3/Q \tag{14}$$

where

$$K_c = (V_s + V_{n\ell}/2)/D^3 \tag{15}$$

Here K_c is a dimensionless constant for a given cyclone design. That is, it depends only upon the relative proportions of the various dimensions. For cyclones which are geometrically similar in all respects it is independent of the size. The magnitude of K_c is an indication of the relative effective volume a given design provides, in which separation of particles from gas may take place.

C. Grade Efficiency

Returning now to equation (6), and replacing t by $t_{res} = K_c D^3/Q$, u_{T_2} by $u_T = Q/ab$, and R_2 by D/2 the following form is obtained:

$$\eta_i = 1 - \exp - 2\left[\frac{(n+1)\ \tau_i Q}{D^3}\left(\frac{8K_c}{K_a^{\,2}K_b^{\,2}}\right)\right]^{\frac{1}{2n+2}} \tag{16}$$

This equation may be displayed in more meaningful form by defining two dimensionless parameters:

$$\Psi_{c_i} = (n+1)\ \tau_i Q/D^3 \tag{17}$$

a cyclone "inertial" parameter reflecting operating conditions within the cyclone, and size of cyclone; and

$$K = 8K_c/K_a^{\,2}K_b^{\,2} \tag{18}$$

a cyclone design number, reflecting only the configuration (shape) of the cyclone. The equation for the grade efficiency is then written

$$\eta_i = 1 - \exp - 2\left[\Psi_{c_i}\ K\right]^{\frac{1}{2n+2}} \tag{19}$$

Equation (19) is equivalent to

$$\eta_i = 1 - \exp - M d_{pi}^{\ N} \tag{20}$$

where

$$M = 2\left[\frac{KQ}{D^3}\ \frac{\rho_p}{18\mu}\ (n+1)\right]^{\frac{1}{2n+2}} \tag{21}$$

and

$$N = 1/(n+1) \tag{22}$$

This is of the Weibull form referred to in equation (36), Chapter 2, and equations (8) and (17), Chapter 3.

The "cut" diameter, for which $\eta_i = 0.50$, will be given by

$$d_{pc} = \left(\frac{0.6931}{M}\right)^{(n+1)} \tag{23}$$

and

$$\eta_i = 1 - \exp - 0.6931\left(\frac{d_p}{d_{p_c}}\right)^{1/n+1} \tag{24}$$

The development and testing of this model are presented by Leith and Licht [6]. Fig. 2 illustrates how it succeeds in reproducing experimental grade-efficiency curves more satisfactorily than other models in the literature. Through the parameter Ψ_{c_i} it correctly indicates that grade-efficiency tends to increase for higher flow rate (Q), denser particles (ρ_p), lower viscosity (μ), and smaller cyclone size (D), for a given cyclone geometry (K). Put in another way, these effects are equivalent to saying that the ratio of particle sizes for equal grade-efficiency is

$$\frac{d_{p_1}}{d_{p_2}} = \left(\frac{Q_2}{Q_1}\cdot\frac{D_1{}^3}{D_2{}^3}\cdot\frac{\rho_{p_2}}{\rho_{p_1}}\cdot\frac{\mu_1}{\mu_2}\right)^{1/2} = \left(\frac{u_{T_2}}{u_{T_1}}\cdot\frac{D_1}{D_2}\cdot\frac{\rho_{p_2}}{\rho_{p_1}}\cdot\frac{\mu_1}{\mu_2}\right)^{1/2} \tag{25}$$

provided the vortex exponent n is essentially constant. This relationship may be used to predict how a grade-efficiency function, determined experimentally under one set of operating conditions, will be transformed under another set of conditions in a geometrically similar cyclone.

There is however one aspect of cyclone performance, namely the effect of grain-loading, which it does not take into account. It is well-known that increased dust loading tends to increase the collection efficiency. This is probably due to agglomeration of particles, coarser particles colliding with and picking up finer ones under the more crowded conditions. The model will conservatively underestimate collection when this occurs. Likewise the collection of very large particles by essentially gravimetric settling may produce greater efficiency than the model predicts. According to Baxter [11], a correction for the effect of grain-loading upon the overall efficiency may be made by

$$\frac{P_{t_1}}{P_{t_2}} = \frac{100 - \eta_{M_i}}{\eta_{M_2}} = \left(\frac{c_{o_2}}{c_{o_1}}\right)^{0.182} \tag{26}$$

D. Effect of Saltation

Equation (19) indicates that as the gas flow Q (or inlet velocity u_T) increases the grade-efficiency will continue to increase and approach 100% I in the limit. However it is known that this is not so, that there is an upper limit to the inlet velocity above which further increase results in a decrease in collection efficiency. This is due to the lack of deposition (or re-entrainment) of particles by the saltation effect, as described by Zenz [12] for horizontal pipelines.

Extending the saltation concept to cyclones, Kalen and Zenz[13] extablished a relationship for the optimum inlet velocity, and the velocity at which particle re-entrainment becomes significant. The saltation velocity may be expressed by an empirical relationship with the inlet velocity

$$u_S = 2.055 \left[\frac{4g\mu\rho_p}{3\rho_f^2}\right]^{1/3} \left[\frac{(b/D)^{0\ 4}}{(1 - b/D)^{1/3}}\right] D^{0.067}\ u_T^{2/3} \tag{27}$$

Units of feet, pounds mass, and seconds must be used in (27). The presence of the cube-root term in this expression relates to $\sqrt[3]{Re_s/C_D}$ used in equation (30), Chapter 4, for the terminal settling velocity correlation.

Kalen and Zenz have shown that the maximum cyclone collection efficiency occurs at $u_T/u_s = 1.25$ and that significant re-entrainment occurs at $u_T/u_s = 1.36$. Koch and Licht [14] found that these criteria are consistent with inlet velocities in the rate of 50 to 90 ft/s in typical cyclones, $(b/D = 0.2)$, which is the commonly accepted range of working values, for particles of density ranging from 1 gm/cm^3 roughly up to 2.5 gm/cm^3.

For $u_T/u_s = 1.25$, (27) reduces to

$$u_T = 16.95\,\frac{4}{3}\,g\left[\frac{\mu\rho_p}{\rho_f^2}\right]\frac{(b/D)^{1.2}}{(1-b/D)}\;D^{0.201} \tag{28}$$

Replacing u_T by Q/ab, and rearranging, an equation for D is obtained:

$$D = 0.0502\left[\frac{Q\rho_f^2}{\mu\rho_p}\,\frac{(1-b/D)}{(a/D)\;(b/D)^{2.2}}\right]^{0.454} \tag{29}$$

This is an empirical equation. The units of ft., lbm., and sec. must be used.

Equation (29) provides a method for selecting the cyclone size D, for a particular configuration, gas flow, and particle and gas properties. It may become the beginning of the design procedure outlined below. Care should be taken however not to use it outside the range of values for which the empirical relationship (27) was developed.

E. Pressure Drop

A number of methods have been proposed to estimate the total pressure drop in the flow of gas through a conventional tangential inlet cyclone. Unfortunately there has not been a definitive study to determine which is the best to use. The total pressure drop is measured between a point in the inlet duct just ahead of the cyclone body, and a point in the exit duct just beyond the top of the cyclone.

It is generally agreed that pressure losses or gains occur at the following places due to the phenomena listed: (a) losses due to frictional flow in the entrance duct; (b) exit losses due to the sudden expansion of the gas stream from the duct into the main cylindrical body; (c) friction losses against the surfaces bounding the gas inside the cyclone; (d) kinetic energy losses due to turbulence within the cyclone; (e) entrance losses due to the sudden contraction of the gas stream at the entrance to the exit duct; (f) static head loss due to difference in elevation between inlet and outlet duct; (g) recovery of energy in the outlet duct; and (h) losses due to frictional flow through the outlet duct. However, the relative importance of each of these items is given different weight in different methods of calculation. Item (d) is generally considered to represent a major source of loss, and some methods assume all other sources to be insignificant by comparison. Items (a), (f), and (h) will obviously be kept very small if the upstream and downstream pressure taps are taken to be very close to the cyclone body.

Most methods agree, however, in expressing total pressure drop in terms of a multiple N_H of "inlet velocity heads." The inlet velocity head is defined as $\rho_f' u_T^2/2g_c$, where $\rho_f' = \rho_f + c(\rho_p - \rho_f)$, c being the volume of dust per unit volume of gas. In terms of inches of water the pressure loss is as given by equation (46), Chapter 3:

$$\Delta P = 0.0030 \, \rho_f' \, u_T^2 \cdot N_H \tag{30}$$

where ρ_f' is in lbm/ft and u_T in ft/sec. The problem is then expressed in terms of finding the value of N_H.

The simplest method is probably that due to Shepherd and Lapple [15] which assumes that item (d) is the only one which needs to be taken into account. It assumes that this loss is proportional to the ratio of area of inlet duct to area of outlet duct and gives

$$N_H = 16\, ab/D_e^2 = 16\, K_a K_G/Ke^2 \tag{31}$$

It is noteworthy that only three of the seven dimension ratios of the cyclone affect N_H according to (31).

More elaborate equations, involving other dimensions of the cyclone, have been proposed by Barth [2], Alexander [7], First [17], TerLinden [17], Miller [18], and Stairmand [19]. Summaries of some of these are given by Strauss [20] without comparing their accuracy. In recent studies by Leith & Mehta [21] and by Gandhi [22] the difficulty arose of finding experimental ΔP data in the literature which were sufficiently complete and detailed to be used as a basis for testing the various methods of calculation. Nevertheless both of these studies concluded (perhaps reluctantly) that one might as well use the simple Shepherd and Lapple method as any other, although in some instances it would produce large errors. Unfortunately it is not clear under what combinations of cyclone dimensions such large errors may be anticipated.

For design purposes here, this approach will be adopted. The fact that all of the dimensions of the cyclone body itself do not affect the pressure drop so calculated, must be watched carefully. The role of N_H in the optimal design of a cyclone will be discussed further below.

III. DESIGN PROCEDURE

A. Standard Configurations

There are a number of cyclone configurations which have been proposed and studied sufficiently that they may be regarded as standards. The dimension ratios of these are tabulated in Table 1, along with the calculated values of ℓ (9), K (18), N_H (31) and the ratio K/N_H. Selection of one of these designs may be regarded as the starting point in a design procedure, and they may also be used as standards for comparison with other proposed configurations.

Equation (19) shows that K is the efficiency parameter, and (31) that N_H is the pressure drop parameter for the cyclone configuration.

Table 1

Cyclone Design Configurations

(Tangential Entry)

Term	Description	"High Efficiency"			"General Purpose"	
		Stairmand [1]	Swift [23]	Lapple [15]	Swift [23]	Peterson & Whitby[5]
D	body diameter	1.0	1.0	1.0	1.0	1.0
a	inlet height	0.5	0.44	0.5	0.5	0.583
b	inlet width	0.2	0.21	0.25	0.25	0.208
S	outlet length	0.5	0.5	0.625	0.6	0.583
D_e	outlet diameter	0.5	0.4	0.5	0.5	0.5
h	cylinder height	1.5	1.4	2.0	1.75	1.333
H	overall height	4.0	3.9	4.0	3.75	3.17
B	dust outlet dia.	0.375	0.4	0.25	0.4	0.5
ℓ	natural length (10)	2.48	2.04	2.30	2.30	2.32
K	equation (18)	551.3	699.2	402.9	381.8	324.3
N_H	equation (31)	6.40	9.24	8.0	8.0	7.76
K/N_H		86.1	75.7	50.4	47.7	44.1
Surf	equation (32)	3.67	3.57	3.78	3.65	3.20
K/N_H Surf		23.5	21.2	13.3	13.1	13.8

The ratio K/N_H is therefore an optimizing parameter. The Table shows that while the Swift design has the highest efficiency, the Stairmand design is optimum by this criterion.

Less efficient shapes may be cheaper, however, by virtue of having less total material in them. This would be indicated by their total surface ratio. A parameter reflecting this may be calculated by adding up the total surface (neglecting the inlet duct) as follows:

$$\frac{\pi}{4}(D^2 - D_e^{\ 2}) + \pi D_e S + \pi D h + \frac{\pi}{2}(D + B)\sqrt{(H-h)^2 + \left(\frac{D-B}{2}\right)^2}$$

This may be put in the form πD^2 (Surf) where

$$\text{Surf} = \frac{1}{4} - \frac{1}{4}\left(\frac{De}{D}\right)^2 + \left(\frac{De}{D}\right)\left(\frac{S}{D}\right) + \frac{h}{D} + \left(\frac{1}{2} + \frac{B}{2D}\right)\sqrt{\left(\frac{H}{D} - \frac{h}{D}\right)^2 + \left(\frac{1}{2} - \frac{B}{2D}\right)^2} \qquad (32)$$

is another dimensionless configuration parameter. Values for Surf are also tabulated.

An overall configuration optimization parameter indicating performance (K) divided by operating cost (N_H) and by initial cost (Surf) may be constructed for comparative purposes. Values of $K/N_H \cdot$ Surf are in the table and again rank the Stairmand design as first choice.

All of the equations developed in the previous section on modelling have only been tested against design similar to those tabulated. These equations should be applied to greatly different configurations only with considerable care and reservations.

B. Steps in Calculations

The design of a cyclone system may be carried out in the following way, assuming specified conditions of:

- Composition and temperature of gas;

- Size-distribution, grain-loading (c_o), and nature of particulate matter;
- Rate of flow of gas stream (Q_o), taken as steady at some average value;
- Desired performance, specified in terms of a maximum allowable emission rate (E).

The overall mass efficiency η_m must be no less than $(1-E/Q_oc_o)$. The usual constraint on the design is

$$\eta_M = \int_o^1 \eta_i \, dG_o \geq (1 - E/Q_oc_o) \tag{33}$$

See Chapter 3, equations (1) and (12). To test this requires a grade-efficiency function η_i.

Steps may be taken in this order: 1. Select a design configuration, for example, the Stairmand, as given in Table 1; 2. Determine D from equation (29); 3. Determine all other dimensions according to the set of configuration ratios; 4. Determine $u_T = Q_o/ab$, and compare this value with that obtained from (28), as a check. If u_T is outside the range of 50-90 ft/s., the reason should be sought. The nature of the particulate matter should also be considered in the light of any relevant experience with it. Values of u_T as high as 150 ft/s have been used with certain materials; 5. Determine ΔP from equation (30); 6. Consider whether D, u_T, and/or ΔP will be excessively large because Q_o is very large. If so, consider using two or more cyclones in parallel; 7. For P cyclones in parallel, repeat steps 2, 4, and 5 using Q_o/P in place of Q_o; 8. Set up equation (20) for the grade-efficiency of each cyclone; calculate M from (21) and N from (22) using (7) for n; 9. Using (20) and the given particle-size-distribution determine η_M using (33), and check the performance constraint.

If the inlet dust happens to have a log-normal size distribution (see Chapter 2, page 31), the calculation of (33) may be done with the aid of a chart prepared by Leith [8], reproduced as Fig. 3. For this chart d_g is the MMD (mass median diameter), and σ_g the geometric standard deviation of the dust. From (23) the cut diameter d_{p_c} is obtained, and N is 1/(n+1). This chart represents the result of the numerical integration of (33) with η_i represented by (20) and G by equation (26), Chapter 2. Refer also to equation (18), Chapter 3.

Note that as the number of cyclones in parallel is increased the value of D for each one will decrease in proportion to $1/P^{0.454}$ and the grade-efficiency for each one will be increased as $P^{0.362}$ affects M in equation (20). M and N will both also be affected somewhat by the value of n, as given by (7). If the required overall efficiency is not reached by a certain design, it may be that using a sufficient number P may cause it to be attained. This gives rise to the system known as multiclones (see below) where N may be 50 or more.

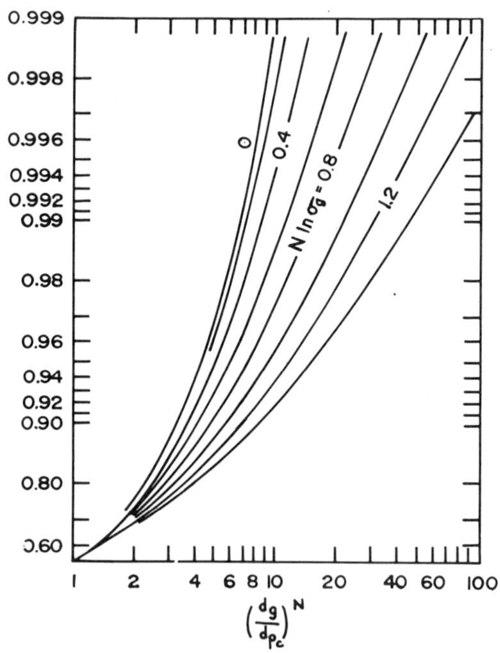

FIG. 3. Overall efficiency for log-normally distributed particulate. Reproduced by permission of Humana Press.

It may be that the required emission limit simply cannot be obtained by any reasonable cyclone design. In that case, the cyclone may still be useful as a primary cleaner to be followed by some other kind of device capable of removing the finer particles with sufficiently high efficiency.

Occasionally it may be desired to design a cyclone to have a specified cut diameter. In this case (23) may be used to determine M, using an estimated value of n, say 0.7. Selecting design configuration (K), the value of Q/D^3 may be calculated from (21). This may be then combined with (29) to find both Q and D individually. The estimated value of n must then be checked against the D value through (7). If necessary the calculations may be repeated using the new value of n. This procedure is discussed further below in connection with the concept of cyclones in series.

C. Practical Considerations

The following are some practical guidelines to consider in contemplating a proposed cyclone design. They are the results of experience and testing.

1. Inlet velocity u_T should be in the range of 50-90 ft/s, generally, although it has gone as high as 150 ft/s in some situations.
2. Dimension a $\lesssim$ S to prevent short circuiting of inlet dust to outlet.
3. Dimension b $\leqq$ (D-De)/2 to avoid too large a ΔP.
4. Dimension H $\geqq$ (S+ℓ) to keep the vortex inside the cyclone; there is no advantage in having H much greater than (S+ℓ).
5. A frequent constraint on the design is the head-room available. This may place a practical limit on H.
6. The angle of the conical base must be steep enough to provide for ready slippage downward of the collected dust; use approximately 7°-8°.
7. Certain trends have been observed by TerLinden [17]: η_M increases with D/De up to about 2.5, then levels, η_M increases with D/De up to

about 10, then levels, η_M peaks at a value of S/De ≃ 1.

8. A practical upper limit to ΔP is about 10" H_2O, but the usual working range is considerably less. For u_T = 50-75 ft/sec., N_H'= 6-9, the value of ΔP from (30) for air ($\rho_f' \simeq 0.075$ lbm/ft^3) will range from 3.4-11.4" H_2O. Hence it is desireable to couple a high value of u_T with a design having a low N_H.

All of the standard designs listed in Table 1 conform well to these guidelines.

Koch and Licht [14] have shown how to apply the design procedure in a quick and preliminary manner using one of the three standard dusts described by Stairmand (see Chapter 3, Table 1). A series of charts are presented to correct for the effect of temperature and of particle density, to select a cyclone diameter and inlet velocity, and to estimate the overall collection efficiency. All of these are in conformity with the equations given above.

The interplay of the effects of all of the operating variables and design parameters upon the cyclone performance is shown by the system analysis chart, Figure 4.

D. Example of Design

The cement dust from a Portland cement kiln is log-normally distributed (as shown in Fig. 9) with an MMD of 12 microns and σ_g = 3.08, having a density of 1.5 gm/cm^3. Operating conditions are: air temperature 250°F, air pressure 1 atm., feed rate to kiln 5 tons/hour, dust rate 230 lbm per ton feed, air rate from kiln 159,600 acfm per ton feed. The Federal emission regulation (for new cement plants) is 0.30 lbm dust per ton feed. Consider whether a cyclone installation might be used to meet this emission limitation.

To meet the emission limitation will require the collection of (230-0.30) x 5 = 1148.5 lbm/hr. with an overall collection efficiency of η_M = (230-0.30)/230 x 100 = 99.78%. While it is unlikely that a cyclone could meet this efficiency, it may be worth while to consider it as a primary cleaner to be followed by a baghouse filter or an

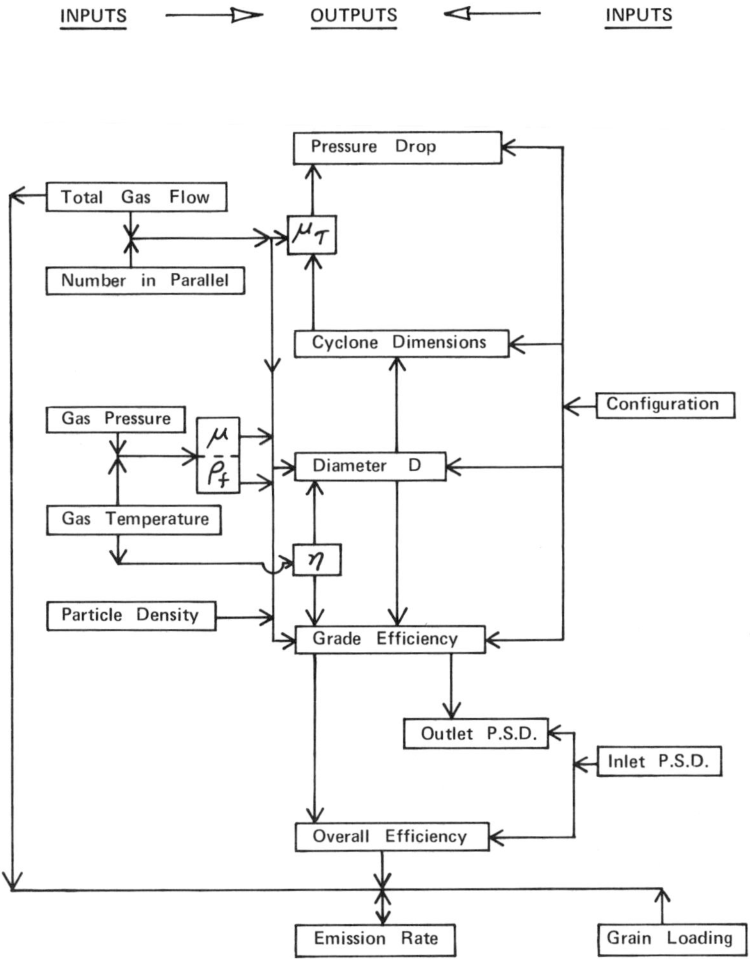

FIG. 4. System analysis of cyclone design.

electrostatic precipitator. The grain loading of 230 x 7000/159600 = 10.1 gr/ft^3, is rather heavy to be a suitable feed for either of these collectors directly.

For purposes of illustration select the design configuration having the largest K, which is the Swift design, Table 1. Using (29) calculate D for a single cyclone:

$$D = 0.0502\left[\frac{221.7 \times (0.0561)^2\ (1-0.21)}{1.512 \times 10^{-5} \times 1.5 \times 62.4 \times 0.44\ (0.21)^{2.2}}\right]^{0.454}$$

$$= \underline{\underline{5.19}} \text{ ft} = \underline{\underline{1.582}} \text{ m}$$

in which Q = 159600 x 5/3600 = 221.7 acfs; K_a = 1/D = 0.44; K_b = b/D = 0.21. At 121 C: ρ_f = 0.0561 lbm/ft^3, μ = 2.25 x 10^{-4} x 0.0672 = 1.512 x 10^{-5} lbm/ft.s. For this design at 250 F (121 C) n = 0.685, equation (7), a = 0.44 x 5.19 = 228 ft., b = 0.21 x 5.19 = 1.09 ft, and u_T = 221.7/2.28 x 1.09 = 89.2 ft/s. This is at a satisfactory value of inlet velocity.

Now set up the grade-efficiency equation (19) in which K = 699.2 (Table 1) and

$$\Psi_{c_i} = 1.685 \times \frac{221.7}{(5.19)^3}\ \frac{1.5\ (d_{p_i} \times 10^{-4})^2}{18 \times 2.25 \times 10^{-4}} = 9.89 \times 10^{-6}\ d_{p_i}^{\ 2}$$

with d_p in microns. Then (19) reads

$$\eta_i\ \ 1\text{-exp-}2\left[9.89 \times 10^{-6} \times 699.2\ d_{p_i}^{\ 2}\right]^{0.297}$$

or in (20) M = 0.4565, N = 0.594. The "cut" diameter (23) will be 2.02 microns. Grade efficiencies may then be calculated and are tabulated as given in Table 2 for P = 1, and plotted in Figure 5.

The overall efficiency may be determined either by the graphical integration method described in Chapter 3, or by the use of Fig. 3. The graphical construction is shown in Figure 6, and gives η_M = 81.7%. In using Fig. 3, the value of the abscissa is $(12/2.02)^{0.594}$ = 2.9 and N $\ell\eta\ \sigma_g$ = 0.594 ℓn 3.08 = 0.67. This chart gives approximately 81% for the overall efficiency.

From (30) with N_H = 9.24 (Table 1), the pressure drop is found to be

$$\Delta P = 0.0030 \times 0.0575 \times (89.2)^2\ s\ 9.24 = 12.7''\ H_2O$$

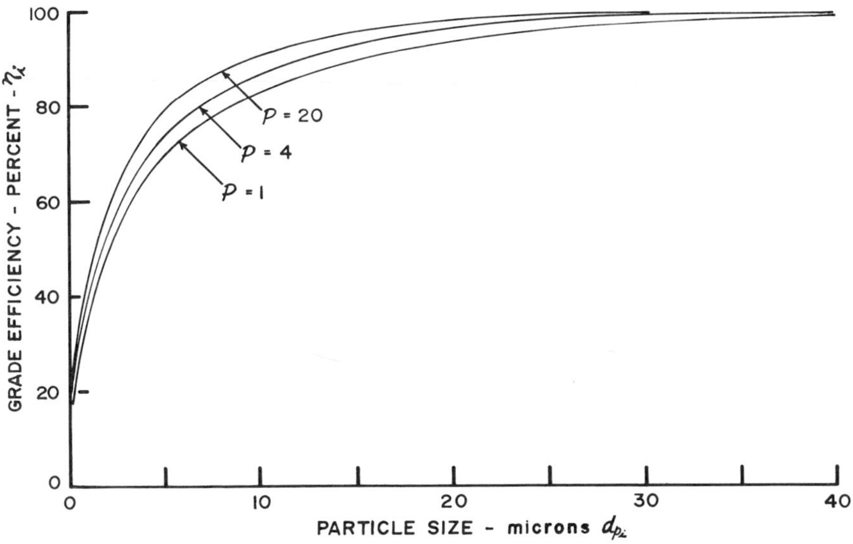

FIG. 5. Grade efficiency for design example.

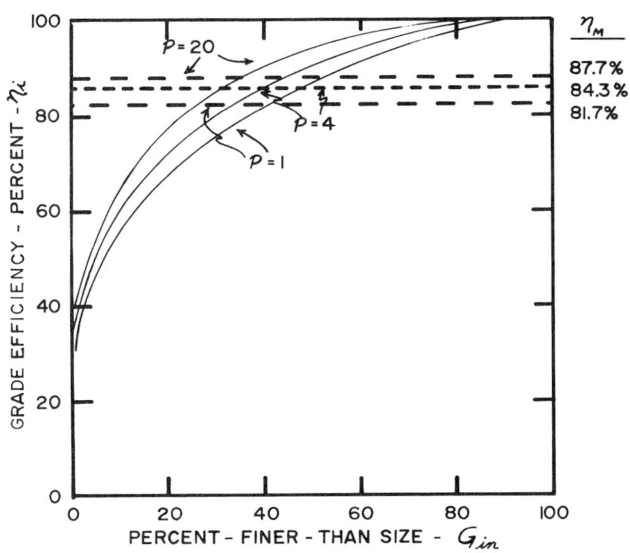

FIG. 6. Graphical integration for overall efficiency.

in which $\rho_f' = 0.0561 + 230/159{,}600 = 0.0575$ lbm/ft^3. This is a very high pressure drop, above the usual range.

Because ΔP is high and η_M is low, consider the use of cyclones in parallel. For illustration, the calculations are repeated with P = 4, and with P = 20. This makes $Q_4 = 221.7/4 = 55.43$ cfs, and Q $Q_{20} = 221.7/20 = 11.09$ cfs respectively. The results are tabulated in Table 2, along with those for P = 1 by way of comparison. Fig. 5 shows the grade efficiency curves, and Fig. 6 the graphical determinations of η_M.

It is clear that the cyclone system can at best serve as a precleaner. The use of 20 small units in parallel will be desireable primarily from the point of view of the reduced pressure drop. For this case then determines what the feed to the secondary cleaner will be. The grain-loading will be reduced to $10.1 \times (1 - 0.87) = 1.31$ gr/acf

The particle size distribution may be determined by the application of equation (16), Chapter 3. This requires the additional graphical integrations in Fig. 6. The inlet and outlet particle size distributions are shown in Fig. 7, and are tabulated in Table 2.

Finally the dimensions of each cyclone in the set of 20 are calculated from the configuration ratios given in Table 1 for the Swift "high efficiency" design.

$$D = 1.33 \text{ ft} = 406 \text{ mm}$$
$$a = 0.44 \times 1.33 = 0.58 \text{ ft} = 178 \text{ mm}$$
$$b = 0.21 \times 1.33 = 0.27 \text{ ft} = 83 \text{ mm}$$
$$S = 0.5 \times 1.33 = 0.67 \text{ ft} = 203 \text{ mm}$$
$$D_e = 0.4 \times 1.33 = 0.53 \text{ ft} = 162 \text{ mm}$$
$$h = 1.4 \times 1.33 = 1.82 \text{ ft} = 555 \text{ mm}$$
$$H = 3.9 \times 1.33 = 5.19 \text{ ft} = 1581 \text{ mm}$$
$$B = 0.4 \times 1.33 = 0.53 \text{ ft} = 162 \text{ mm}$$

This design problem will be treated further in Chapters 7 and 9, when the design of the secondary cleaners will be discussed.

IV. SPECIAL DESIGNS AND SYSTEMS

There are significant limitations upon the amount of particulate collection which may be obtained with a single reverse-flow cyclone, as has been pointed out. More complex designs and systems may be considered to achieve better performance. Among these may be listed: (a) two or more cyclones in parallel; (b) two or more cyclones in series; (c) recycling portion of emission stream; (d) skimming off portion of uncollected dust in emission stream; (e) use of secondary air flow. Each of these will be discussed briefly.

As the above design example indicates, the use of P cyclones in parallel produces improved performance, and it is a customary practice. Grade efficiencies are increased and pressure drop is decreased when they are designed in the way described. The total cost is undoubtedly increased however, partly because the total material, represented by ($\pi D^2 \cdot$ Surf $\cdot$ P) in Table 2, increases and partly because of the increased complexity of duct work and installation. Furthermore care must be taken to assure that each cyclone in the set is operating in the same manner as all the others. If one or more has a different pressure drop (due to wear, or partial blockage etc.) then they no longer operate truly in parallel and the flow rates are not the same in each. There may even be secondary circulations if all the cyclones are connected to a common dust hopper as in a multiclone. There effects can seriously affect the overall performance.

Placing two or three cyclones in series downstream, so that the emission from one becomes the feed to the next, is also practiced. The principal reason for this practice is usually to recover separately two or more different size fractions of the product. The design may be based upon the alternate procedure, mentioned above, of specifying a desired cut diameter for each fraction. In this case the downstream cyclone would probably be of a more efficient configuration and/or smaller size than the upstream one. This is intended to produce increased recovery of the finer sized particles which have escaped the first unit. It is generally not possible to obtain a real sharp separation of particle sizes in the products. Note that since each unit must handle the total flow the overall pressure drop is greatly increased.

Recycling a portion of the outlet stream, back into the inlet of a single unit (or set of parallel units), might also be contemplated as a method of increasing overall efficiency. It can do so, but will require a much larger cyclone and a greater power consumption to treat a given volume flow of gas.

Another, relatively new, recycling design involves arranging for that portion of exit gas near the inside wall of the duct D_e to be removed ("skimmed off") through a Venturi-shaped annular insert in the top of D_e. This is thought to remove the majority of the uncollected particles which are presumed to be exiting near the outlet tube walls. The skimmed-off stream is recycled by a secondary fan into the inlet duct of the cyclone. At least two commerical versions of this are on the market and both claim very high grade efficiencies with "cut" diameters below 1 micron. No theoretical analyses have been published. It would seem that the action would be like two units in series, with the added function of a very high inlet velocity due to the added power in the circulated stream.

Finally may be mentioned what has been variously called the "rotary flow," or "straight-through with reverse flow" cyclone [24, 25]. It consists of a cylindrical body into which two separate streams of gas are introduced. The primary flow, the gas to be cleaned enters at the bottom either tangentially or through a central set of turning vanes. A secondary flow, which may be either a fresh stream or a recirculated portion of the emission gas, is introduced tangentially at the top. As particulates in the primary flow are thrown outwards toward the wall they are swept downward to the collection hopper by the secondary flow. The particles are dropped into the hopper when the secondary flow reverses and joins the primary flow. Clean gas is emitted through a central-core at the top. This design is supposed to give a much increased grade-efficiency by preventing reentrainment and offseting the saltation effect. Other advantages claimed for it are: relative insensitivity to grain loading up to large values, reduced abrasion of the walls, and the ability to cope with hot gases (which are cooled by the secondary flow) without using special heat-resistant materials of construction.

Published grade-efficiency curves [24] do indeed show very high values with cut diameters in the range of 0.5 microns. They are of the same general shape as the experimental curves shown in Fig. 2, and can be fitted empirically by equation (20), although the physical model leading to (20) obviously does not apply. Ciliberti and Lancaster [26] studied the performance experimentally and developed a model which gives quite different results, however. Their model predicts, and their experiments confirm, that the grade-efficiency drops to zero at a minimum particle size in the range of 1-3 microns, with cut diameters greater than 1 micron.

Since the secondary gas flow is roughly of the same order as the primary, it is evident that much greater pressure loss and power requirements will obtain for this design.

Table 2
Results of Design Example

Quantity		Source on Equation	Number of Cyclones in Parallel		
			P=1	P=4	P=20
Q	acfs		221.7	55.43	11.09
D	ft.	(29)	5.19	2.77	1.33
u_T	f/s		89.2	78.3	67.9
n	-	(7)	0.685	0.618	0.548
ΔP	"H_2O	(30)	12.7	9.8	7.3
$K\psi_{c_i}/d_p^2$	μm^{-2}	(17, 18)	6.92×10^{-3}	1.09×10^{-2}	1.89×10^{-2}
M	*	(21)	0.4565	0.4951	0.5550
N	-	(22)	0.594	0.618	0.646
d_{Pc}	μm	(23)	2.02	1.72	1.41
η_M	%	Fig. 6	81%	84%	87%
$P(\pi D^2)$Surf	ft^2		302	344	397

Table 2 (continued)

d_p	μm	(20)	Grade Efficiency - %		
	1		36.7	39.1	42.6
	3		58.4	62.4	67.7
	5		69.5	73.8	79.2
	10		83.3	87.2	91.4
	20		93.3	95.7	97.8
	30		96.8	98.3	99.3
	40		98.3	99.2	99.8
	50		99.1	99.6	99.9
	60		99.4	99.8	99.9
η_M		Fig. 6	81.7	84.3	87.7

Results of Design Example-Outlet Particle Size Distribution

dp-μm	G_{in}	"Area"$^{P=1}$ Fig.6	G_{out}	"Area"$^{P=4}$ Fig. 6	G_{out}	"Area"$^{P=20}$ Fig. 6	G_{out}
2.8	0.10	0.0456	0.300	0.490	0.325	0.0534	0.380
4.7	0.20	0.1078	0.505	0.1154	0.539	0.1248	0.613
6.7	0.30	0.1794	0.661	0.1909	0.695	0.1060	0.767
9.1	0.40	0.2579	0.778	0.2736	0.806	0.2934	0.870
12.0	0.50	0.3419	0.866	0.3615	0.883	0.3854	0.935
16.2	0.60	0.4306	0.928	0.4536	0.933	0.4809	0.972
22.0	0.70	0.5233	0.968	0.5462	0.980	0.5787	0.989
31.5	0.80	0.6195	0.989	0.6437	0.996	0.6777	0.998
52	0.90	0.7177	0.998	0.7431	0.999	0.7774	0.999
	1	0.8174	1	0.8431	1	0.8774	1
P		0.183		0.157		0.123	

$$G_{out} = \frac{1}{P}\left[G_{in} - \int_0^{G_{in}} \eta_i dG_{in}\right]$$

$$\text{"Area"} = \int_0^{G_{in}} \eta_i dG$$ – refers to Fig. 8

$$\eta_M = \int_0^1 \eta_i dG_{in}$$

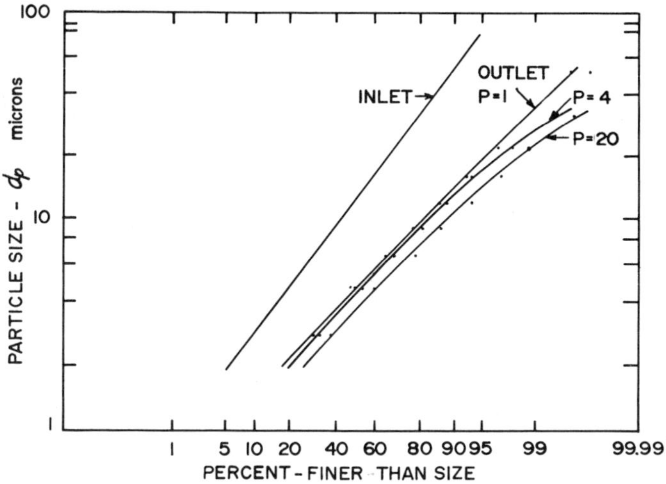

FIG. 7. Particle size distributions for inlet and outlet of cyclone.

REFERENCES

1. C.J. Stairmand, *Trans. Inst. Chem. Engrs.*, 29: 356 (1951).

2. W. Barth, *Brennstoff-Warme-Kraft*, 8: 1 (1956).

3. C.E. Lapple, in *Air Pollution Engineering Manual*, USEPA AP-40, 2nd ed., 1973.

4. W.T. Sproull, *Air Pollution and Its Control*, Exposition Press, Jericho, N.Y., 1970.

5. C.M. Peterson and K.T. Whitby, *A.S.H.R.A.E.J.*, 42: (1965).

6. D. Leith and W. Licht, *A.I.Ch.E. Sympos. Ser.* 68: No. 126, 196 (1972).

7. R. McK. Alexander, *Proc. Austral. Inst. of Min. and Met. (N.S.)*, 152/3: 202 (1949).

8. D. Leith in *Handbook of Pollution Abatement Engineering*, Humana Press, Clifton, N.J. in press 1978.

9. A.J. Terlinden, *Proc. Inst. Mech. Engin.*, 160: 233 (1949).

10. M. Aslami and W. Licht, *A.I.Ch.E. Symp. Series* No. 175: 74, 18 (1978).

11. W.A. Baxter in *Air Pollution-Vol IV, 3rd Ed.*, A. Sterr ed., Chap. 3, Academic Press, New York, 1977.

12. F.A. Zenz, *Ind. Eng. Chem. Fund.*, 3: 65 (1964).

13. B. Kalen and F.A. Zenz, *A.I.Ch.E. Sympos. Ser.*, 70: No. 137, 388 (1974).

14. W. Koch and W. Licht, *Chem. Engin.*, 84: 80 (Nov. 4, 1977).

15. G.B. Shepherd and C.E. Lapple, *Ind. Eng. Chem.*, 31: 972 (1939).

16. M.W. First, *Amer. Soc. Mech. Engrs.*, Paper 49-A-127 (1949).

17. A.J. TerLinden, *Toindustrie-Zeitung, 22 (iii):* 49 (1953).

18. F. Miller and Lissman, *Amer. Soc. Med. Engrs.*, Paper, New York (Dec. 1940).

19. C.J. Stairmand, *Engineering*, 409 (Oct. 1949).

20. W. Strauss, *Industrial Gas Cleaning*, 2nd ed., Pergamon Press, Oxford, (1975).

21. D. Leigh and D. Mehta, *Atmos. Environ.*, 7: 527 (1973).

22. M. Gandhi, M.S. Project in Chem. Eng., Univ. of Cincinnati, (1972).

23. P. Swift, *Steam Heating Eng'g.*, 38: 453 (1969).

24. W. Strauss, *Industrial Gas Cleaning*, 2nd ed., pp. 263-4, Pergamon Press, Oxford, 1975.

25. K.R. Schmidt, *Staub*, 23: 491 (1963).

26. D.F. Ciliberti and B.W. Lancaster, *A.I.Ch.E. Journ.*, 22: 394, 1150 (1976).

CHAPTER 7

ELECTROSTATIC PRECIPITATION

I. THE ELECTROSTATIC PRECIPITATION SYSTEM

An electrostatic precipitator collects particles by means of an electrostatic force. The aerosol particles are first given a charge and then passed through an electrostatic field. The collecting force on a particle is the product of the charge (q) and the potential gradient (E) of the field.

$$F_E = q \cdot E$$

By using a very large value of E (typically, of the order of 2000 V/cm) it is possible to establish a fairly large force even upon a very small particle. The collection efficiency for very small particles may thus be made very high, and the principle finds its greatest application in the collection of them.

To implement the concept it is necessary to provide: (a) a method of giving a charge to the particles flowing in the gas stream; (b) establishment of an electrostatic field through which the gas stream may flow at right angles; (c) a method of collecting the particles after they have migrated across the stream; this involves removing the charge once the particle is essentially out of the stream.

These steps are accomplished in a device consisting essentially of two electrodes with space between them for the gas to flow. One electrode is a high-potential wire, freely suspended, called the corona-discharge or "active" electrode. The other is a grounded extended surface called the collecting or "passive" electrode. One of two geometrical arrangements is usually used: the collecting electrode is a vertical hollow cylinder with the wire suspended coaxially within it; or, the collecting electrodes are vertical parallel flat plates with several wires suspended midway between the plates at uniform intervals of spacing. The device is one of the "containment zone" or "surface" collectors described in Chapter 3.

When a sufficiently great potential difference is set up between the electrodes, a corona discharge forms about the wires. In this corona, molecules of gas (principally O_2) become ionized with charges of the same sign as the wire. They migrate across the space between the electrodes under the influence of the applied electrostatic field. In doing so they collide with and attach themselves to the aerosol particles which thus acquire charge and also migrate toward the grounded surface electrode. When the particles strike the grounded electrode they lose charge and adhere to this surface. The layer of solid particles which thus accumulates is removed by vibrating (rapping) or flushing the electrode, and drops or slides down into a collecting hopper. Thus steps (a), (b) and (c) are completed.

If the potential of the wire is made too high, there will be a breakdown and sparking over the gas stream to the grounded electrode. This, of course, destroys the collecting field. However, since it is desirable to operate as close to the limiting potential as possible, sparking is permitted to occur in controlled fashion in industrial operations. The potential of the corona wire may be either positive or negative, but in industrial cleaners it is usually made negative because a higher potential can be obtained before sparking occurs. The sparking may cause the formation of undesirable gaseous products such as ozone or oxides of nitrogen.

There is a modification of the basic design which is known as the two-stage precipitator. In this the corona-discharge particle charging is carried out in a first stage or charging zone, with the particle collection taking place in a secondary zone in which there is merely a high voltage non-discharging electrode to maintain the field.

Complete consideration of the design and performance of electrostatic precipitators must of course include the purely electrical aspects of it. These involve the phenomena of corona discharge and ion-generation as well as the relationships between voltage applied, current, and field-strength produced. Since the present discussion is devoted primarily to the particle collection process itself, details of the electrical aspects will not be presented. The reader is referred to the excellent reviews by White [1], Robinson [2], or Oglesby [3] for this type of information.

The complete electrostatic precipitator system is undoubtedly one of the more complex types of collectors. There are a number of sub-systems within it, and interaction between the electrical and particle aspects of it is complex. Figure 1 modified from Nichols and Oglesby [6], is an attempt to show the entire system and the interrelationships within it. The heavy lines are intended to link the principal steps involved in particle collection. The lighter lines represent the subsidiary interactions which also influence the process.

II. MODELLING THE SYSTEM

A. General Aspects

Modelling of such a complex system in its entirety is obviously very difficult and has not been accomplished. Piecemeal-consideration of various parts of the model, necessarily in more or less approximate fashion, has been presented in the earlier literature. Nichols' and Oglesby's attempt to deal with the entire system [6], based upon the exhaustive study made by the Southern Research Institute [3, 4] is perhaps foremost among a number of new approaches [7, 8, 9, 10].

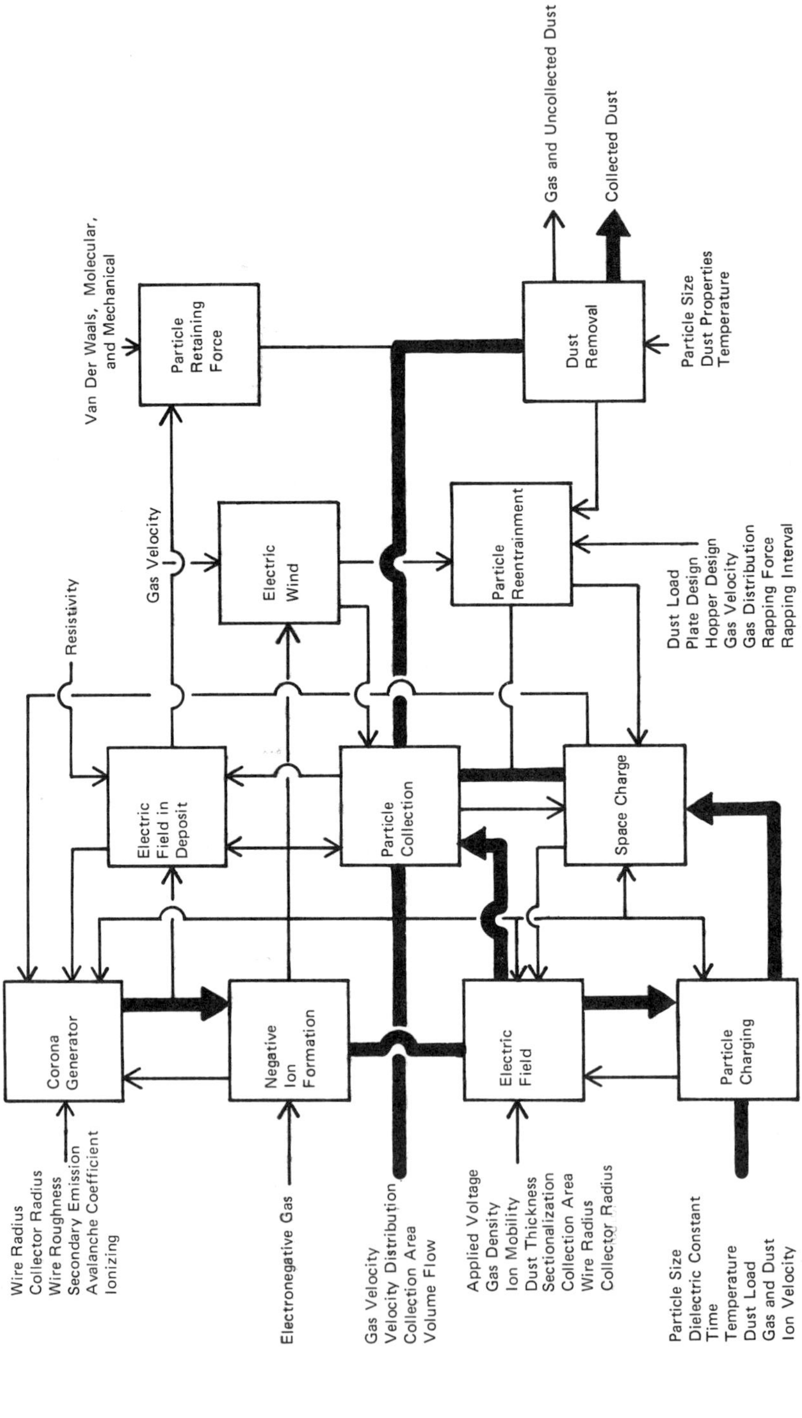

FIG. 1. Electrostatic precipitator system model [6]. Reproduced by permission from Southern Research Institute.

Until recently there have not been many detailed reports of actual grade-efficiency relationships available for use in modelling studies. Those which have been published tend to have the general appearance of those shown in Fig. 2 and 6, Chapter 3, except that the efficiency values are much higher in the small-particle size ranges. An irrigated precipitator (plates cleaned by flushing) tends to give even higher efficiencies than a dry one (plates cleaned by rapping) due no doubt to reduction of re-entrainment losses. Recent studies sponsored by the E.P.A. provide more detailed information in the sub-micron range of greatest interest. Fig. 2 reveals a behavior which was not previously documented, namely a range of minimum grade-efficiency near 0.2 to 0.8 microns [11].

Pressure drop through the precipitator is relatively small (typically less than 0.5" H_2O) since the gas flow is rather unimpeded. Under usual conditions (velocity 4-8 ft/s., plate spacing 8"-12") in a parallel plate device, Re will be of the order of 10^4. A high degree of turbulent flow is therefore common in large scale units. Transport of fine (submicron) aerosol particles by turbulent diffusion may easily become more important than by the electrostatic force, except in the boundary layer adjacent to the collecting electrode. Coarser particles may have sufficient inertia to be relatively unaffected by the fluid turbulence. Thus a lateral-mix model (see Chapter 3) may apply to the fine particles, but not so well to the coarse ones. As collection proceeds this could lead to the central portion of the gas stream around the corona electrode becoming depleted of particles toward the downstream end of the collector. The result would be a very non-uniform distribution of dust concentration in cross-section normal to the flow.

The gas ions also stream from active toward passive electrodes under the influence of the field potential. This sets up a cross-current of gas flow referred to as the "electric wing." It obviously will have an effect upon particles distribution, collection, and re-entrainment.

It cannot be assumed always that once a particle has reached the collecting electrode, it is removed and no longer influences the col-

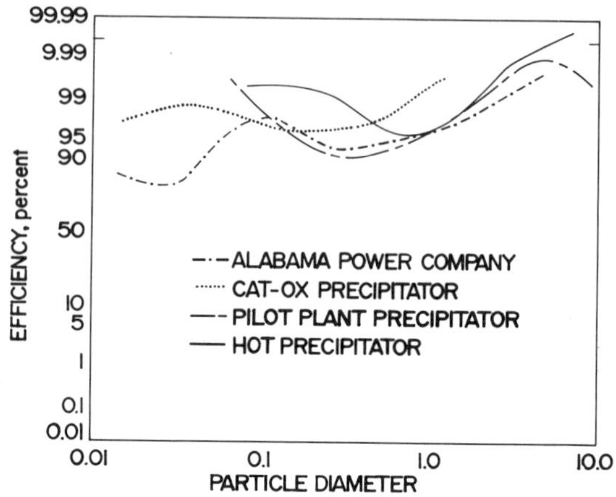

FIG. 2. Measured grade efficiency for precipitators [11]. Reproduced by permission from Chemical Engineering Progress.

lecting process. If the particle material is a good conductor it may lose its charge at once and easily become re-entrained in the gas stream. If it is composed of a poorly conducting material, it may tend to cling to the electrode and build up an insulating layer upon it. This hinders the dust collection and may lead to a sparking discharge, indeed to a "reverse corona" effect. There is an optimum range, expressed in terms of particle electrical resistivity running between about 10^4 and 10^{10} ohm-cm., wherein the discharge proceeds smoothly. Electrical resistivity is affected strongly by the temperature and composition of the gas in the system.

Nor can the importance of the rather simple geometry of the device be overlooked. Uniformity of electrical field is highly desireable and depends, among other factors, upon uniformity of electrode spacing. If the hanging wire electrode is not quite parallel to the collecting surface, then the potential gradient E becomes higher wherever the spacing is smaller. Since the vertical distance is of the order of several feet, while the electrode spacing is a matter of inches, it is easy to see how a very small departure from "plumb" can create a significant variation in E over the collecting zone.

Uniformity of gas velocity distribution throughout is also very important. Regions of high velocity adversely affect the collection to disproportionate degree, as will be shown later.

One of the advantageous features of an electrostatic precipitator is that it may be built to treat hot gas streams. It has thus come into use for collecting particles from hot (up to perhaps 800-900 F) stack gases, such as fly-ash arising from coal combustion. In modelling calculations this will require an accurate knowledge of the values of all the relevant physical properties (e.g., gas viscosity, particle resistivity, ion mobility, etc.) at the temperature in question. Values in use for ordinary temperature may be seriously in error.

These then are some of the important considerations to be taken into account in the modelling. They are shown in Fig. 1 along with other details. Because of the complexity of the total system, it is helpful to consider first a rather elementary or first-order approximation model and then to see how it requires modification.

B. Elementary Model for Particle Collection

An elementary, over-simplified, model of the collecting process may be developed beginning with the calculation of a collecting velocity due to the electrostatic force. This was begun, for Stokes Law particles, in Chapter 4, example 6, by setting up the equations for the trajectory of a particle having charge q in field of potential gradient E. Based upon equation (53) there, it was shown that the distance travelled in the direction of the field is:

$$z = \frac{qE\rho_p d_p}{54\pi\mu^2}\left[\frac{t}{\tau} - (1-e^{-t/\tau})\right]$$

This indicates that after a period of time $t \simeq 5\tau$, a steady state is attained with velocity $u_{E_s} \equiv w = qE/3\pi d_p\mu$. This is called the "steady-state particle migration velocity," customarily denoted by w. It assumes that the particle has acquired and keeps a steady charge q* and that the potential gradient is constant. In case the particle is small enough to require the Cunningham factor correction to Stokes' Law it would become

$$w = q^*EC/3\pi d_p \mu \tag{1}$$

Sample calculations of Re_p at typical precipitacor conditions indicate that Stokes' Law will be valid for particles up to 60 microns or so. As pointed out by Reynolds and Theodore [8] there may be some applications which fall outside this range.

The following assumptions are now made to describe the model:

(1) The particles are considered to acquire the full charge q* immediately on introduction into the collection zone. (The question of calculating the value of q* will be considered later).

(2) There is a collection layer of thickness, δ, on the discharge electrode in which the flow is laminar, but in the main core of the stream it is turbulent.

(3) The thickness, δ, of the collection layer is defined as the distance away from the collection electrode to the point where the linear velocity of the laminar flow gas stream, down the precipitator, is equal to w.

(4) In the collection layer the particle always moves cross-stream at its migration velocity, which is unaffected by the velocity of the gas stream, and determined by the value of E in this layer region.

(5) In the turbulent core of the gas stream, the fluid velocity is taken as uniform at the average gas velocity v_o. (The question of non-uniform velocity will be dealt with below.)

(6) Turbulence and diffusion forces keep the particles uniformly distributed at concentration n_i across the central core in accord with (2), and the effect of migration velocity is negligible in this zone; the lateral-mix model described in Chapter 3 is assumed. (This may limit the applicability of the model to finer particles.)

(7) Dust particles are sufficiently separated that their mutual repulsion can be neglected.

(8) There are no disturbing effects such as erosion, reentrainment, uneven gas flow distribution, or back-corona present.

During a period of time Δt, defined by $\Delta t = \delta/w_i$, all particles of ith grade within the collection layer will move to the wall and, by definition, be collected. During this period of time the gas will move a distance $\Delta \ell = v_o \Delta t$. The number of particles thus removed from the volume of gas given by $(S\delta\Delta \ell)$, where S is the perimeter dimension of the collecting surface, will be

$$n_i \; S \; \delta \; \Delta\ell = n_i \; S \; \delta \; v_o \; \Delta t$$

At the same time, the total number of particles in the corresponding volume of the precipitator, having cross-sectional area A, will be $n_i \, A \cdot \Delta\ell$. Thus the fractional increment of particles removed is

$$\frac{dn_i}{\eta_i} = - \frac{S}{A} \frac{W_i}{v_o} \, d\ell$$

Then integrating, under all of the assumptions, over the length, L, of the collecting surface:

$$\int_{n_{oi}}^{n_i} \frac{dn_i}{n} = - \frac{S}{A} \frac{w_i}{v_o} \int_0^L d\ell$$

gives the efficiency of collection as

$$\eta_i = 1 - \exp\,(- SL \; w_i/Av_o) \tag{2}$$

Note that SL equals the total surface of collecting electrode, and that $A \, v_o$ equals the volumetric rate of gas flow.

In the case of a parallel plate precipitator A = bS, where b = half-width of spacing between plates, and in the case of a tubular precipitator $A = \pi D^2/4$ where D is the diameter of the tube. Thus (2) may be written

$$P_i = \exp - \frac{w_i}{v_o} \frac{L}{b} \qquad \text{(plate)} \tag{2a}$$

or

$$P_i = \exp - 4 \frac{w_i}{v_o} \frac{L}{D} \qquad \text{(tubular)} \tag{2b}$$

Since w_i depends upon d_{pi}, equation (2) gives a grade-efficiency relationship. In the Stokes Law range (1) may be substituted for w_i, giving

$$\eta_i = 1 - \exp\left(- \frac{SL}{Q} \frac{C_i q_i^* E}{3\pi\mu d_{pi}} \right) \tag{3}$$

A relationship for q_i^* is required to complete this model.

C. Particle Charging

To express (3) in an explicit grade-efficiency form, a relationship between q_i^*, and d_{pi} must be known. The development of this may be summarized as follows. For detailed derivation see Crawford [12].

Particles acquire charges by two distinct mechanisms referred to as (a) ion bombardment, and (b) ion diffusion [2]. These are also known respectively as field charging and diffusion charging. They are due (a) to motion of the ions directionally along the lines of flux, set up by the applied electric field and (b) to random thermal motion of ions and particles (see Brownian motion, Chapter 4).

1. Field Charging

The lines of flux in the neighborhood of an uncharged particle are drawn into it. Hence ions travelling along them readily collide with and adhere to the particle. As charges build up however, the flux lines are gradually repelled until finally they no longer bring ions into contact with the particles. This form of charging then stops and the particle is said to have acquired a saturation or maximum charge from ion bombardment. The charge builds up with time according to [2]

$$q_i^* = q^*_{sat_i}\left(\frac{1}{1+t_o/t}\right) \qquad (4)$$

where $t_o = 4\ \varepsilon_o/NeK$, a time constant independent of d_D, and

$$q^*_{sat_i} = \pi\left(\frac{3\varepsilon}{\varepsilon+2}\right)\ \varepsilon_o E_o d^2_{p_i} \qquad (5)$$

Here N = ion concentration, number/m^3; K = ion mobility, velocity per unit field strength, $m^2/s{\cdot}V$. The other symbols are as given in Chapter 4.

For practical purposes q^*_{sat} may be taken as attained (99%) at $t = 99\ t_o$. This is usually a very short period of time because $N \simeq 10^{15}$ ions/m^3, $K \simeq 10^{-4}$ m/s V, $\varepsilon_o = 8.85 \times 10^{-12}$ coulombs^2/Nm^2, and $e = 1.61 \times 10^{-19}$ coulombs. Therefore

$$t_o \simeq \frac{4 \times 8.85 \times 10^{-12}}{10^{15} \times 1.61 \times 10^{-19} \times 10^{-4}} \simeq 2.2 \times 10^{-4}\ s$$

and $99\ t_o \simeq 0.022$ s. At a maximum gas velocity of say 8 ft/s, particles of all sizes would have acquired the saturation charge after traveling at most only about 2" into the charging zone.

2. Diffusion Charging

In the case of diffusion charging, collisions between particles and ions resulting from the random thermal motion continue as long as the particles are in the ion field. No saturation charge exists. The diffusion charge builds up according to

$$q^*_i = \frac{2\pi\varepsilon_0 kTd_{p_i}}{e} \ln\left[1 + \frac{e^2Nd_{p_i}t}{2\varepsilon_0\sqrt{2\pi mkT}}\right] \tag{6}$$

assuming $q^*_i = 0$ when $t = 0$. Here m = mass of an ion, and k is the Boltzman constant [2].

The relative importance of the two methods of charging depends primarily upon the particle size. For particles roughly 0.5 micron and larger, field charging is predominant. Below that size diffusion charging becomes increasingly important as the values of $q^*_{sat_i}$ become very small. Most authors [2, 7, 12] recommend simply adding (5) to (6) to obtain the total charge on a particle as a function of the time it has been exposed to the charging field.

For a set of typical operating conditions: $\varepsilon = 5$, $E_o = 3 \times 10^6$ V/m $T = 300°K$, $N = 2 \times 10^{15}$ ions/m^3, $m = 5.3 \times 10^{-26}$ kg (mass of an O_2 ion), Crawford [12] has calculated that the total charge obtained by adding (5) and (6) will be

$$q^*_{Ti} = 178.7 \times 10^{-6} d_{p_i}^2 + 1.438 \times 10^{-12} d_{p_i} \ln(1 + 7.79 \times 10^{10} t d_{p_i})$$

in which q^* is in coulombs, d_{p_i} in meters, and t in seconds. The first term represents the value of $q^*_{sat_i}$ hence

$$q^*_i/q^*_{sat_i} = 1 + \frac{8.047 \times 10^{-9}}{d_{p_i}} \ln(1 + 7.79 \times 10^{10}\ t d_{p_i})$$

From these equations the values tabulated on the next page have been calculated. (Note that $t_o = 5.03 \times 10^{-4}$ s. and $99\ t_o \simeq 0.05$ sec.) The trends mentioned above are clearly indicated. Looking to the application of such calculations in (3), values of the Cunningham slip factor C_i are also included.

d_{p_i}-μm	0.05	0.08	0.1	0.3	0.5	1.0
$q^*_{sat_i}$ -[a]	0.45	1.14	1.79	16.1	44.7	178.7
			Values of $(q^*/q^*_{sat_i})$			
At t:						
0.1 sec	1.96	1.65	1.54	1.21	1.13	1.07
1 sec	2.33	1.88	1.72	1.27	1.17	1.09
10 sec	2.70	2.11	1.91	1.33	1.21	1.11
C_i	4.34	3.09	2.67	1.56	1.33	1.17
	Values of (Cq^*_i/d_{p_i}) (q^*_i/q^*_{sat}) at t = 1 sec. [b]					
	91.0	82.8	82.2	106	139	224

[a] Units of $q^*_{sat_i}$ are coulombs x 10^{18}
[b] Units are coul x 10^{18} x μm^{-1}

D. Grade-Efficiency Equations

Equation (3) with (5) plus (6) substituted for q^*_i becomes the explicit grade-efficiency relationship. Since q^*_i is of the form

$$q^*_i = ad_{p_i}{}^2 + bd_{p_i} \ln (1 + ctd_{p_i})$$

(3) will become the form

$$\eta_i = 1 - \exp - KC_i \, [ad_{p_i} + b \ln (1 + ctd_{p_i})] \tag{7}$$

For the coarser particle size range the ln term is negligible and $C_i \approx 1$, so that

$$\eta_i = 1 - \exp - \frac{SL}{Q} \left(\frac{\varepsilon}{\varepsilon + 2}\right) \frac{\varepsilon_o E_o E d_{p_i}}{\mu} \tag{8}$$

This is the form recommended by White [1] for estimating precipitator performance. Note that it is of the form of equation (17) Chapter 3, with N = 1.

In a two-stage precipitator E_o (for the charging stage) and E (for the collecting stage) are independently controlled and may very well be quite different. In a single-stage precipitator they are often assumed to be equal. However, assumption (4) in the list above requires that E be taken in the vicinity of the collecting plate; E does not necessarily equal E_o. Gooch and Francis [7] recommend taking E_o at the average value between discharge and collecting electrodes, and determining E by the numerical method of Leutert and Bohlen [13]. This is further detailed by Gooch and McDonald [14].

Equation (8) agrees with the form of experimental grade-efficiency determinations on spherical particles of uniform composition in which the assumptions underlying (2) are reasonably met. However, it cannot account for the results of tests such as are cited in Fig. 3.

To account for the behavior of submicron particles the complete form of (7) must be used. Since the ℓn term is time-dependent, it will be necessary to specify the period of charging time to use in order to make grade-efficiency calculations. This may be done by using the function of d_{p_i} in the form $(Cq^*_{sat_i}/d_{p_i})(q^*_i/q^*_{sat_i})_t$. These values are also cited in the above tabulation taking t = 1 s. The combination of factors produces a minimum in this value in the neighborhood of particle size of 0.1 micron. This is consistent with the experimental results depicted in Fig. 2. Other factors, not included in the model may also influence these tests.

Another representation of the total charge has been used by Feldman [18] to account for the behavior of the sub-micron particles. According to Cochet [19, 20] the combined effect of field charging and diffusion charging leads to

$$q^*_i = [(1 + \frac{2\lambda}{d_{p_i}})^2 + \frac{2}{(1 + 2\lambda|d_{p_i})} \frac{\varepsilon - 1}{\varepsilon + 2}] \pi \varepsilon_o E_o d^2_{p_i} \frac{t}{t+t_o} \qquad (9)$$

in which λ represents the mean-free path of the gas (taken as 10^{-7} m in room air) and t_o is the time-constant given in (4). Taking $(\varepsilon-1)/(\varepsilon+2) \approx 1$, and $t >> t_o$, Feldman incorporated (9) into (3) to obtain

$$\eta_i = 1 - \exp - \frac{SL}{Q} \frac{\varepsilon_o E_o E}{3\pi} (C_i A_i d_{pi}) \tag{10}$$

as a grade-efficiency equation to be compared with (7). Here

$$A_i = (1 + 2\lambda/d_{pi})^2 + 2/(1 + 2\lambda/d_{pi})$$

Feldman showed through sample calculation that the size-dependent term $(C_i A_i d_{pi})$ in (10) goes through a minimum value in the range of d_{pi} between 0.2 and 0.8 μm.

E. Deutsch-Anderson Equation

1. Precipitation Rate Parameter

Equation (2) as it stands, without refinements, is known as the Deutsch-Anderson equation and has been used as the basis for much work on precipitators. While it can be taken as a grade-efficiency equation, it has been the practice to use it for overall efficiency. This is done by replacing w_i with a so-called "effective" migration velocity, or more properly "precipitation rate parameter" w_e. This quantity is taken to represent the collection behavior of an entire dust of a certain kind and under a certain set of operating conditions. It is determined by back calculation from experimental data and so is really a performance parameter.

Quite a bit of such data is available (although much of it is proprietary) [1, 2, 3, 4] for a variety of applications. Values of w_e are related to kind of dust, gas velocity, length of surface L, particle size, design of electrodes, rapping methods, sparking routines, etc. They are used as a basis of empirical precipitator design usually in the form:

$$\eta_M = 1 - \exp\left(-\frac{SL}{Q} w_e\right) \tag{11}$$

Examination of such data indicates that the assumptions underlying (2) may not always be met, especially number (8). The influence upon the model of such factors as uneven gas flow distribution, particle resistivity, diffusional effects, etc. must be considered although for the most part it cannot be done quantitatively. This is treated briefly below.

F. Overall Efficiency

The calculation of overall efficiency from grade-efficiency must be done in accord with the general principles presented in Chapter 3. Thus using (2) or (10), $P = 1 - \eta_M$ is

$$P = e^{\frac{-SL}{Q} w_e} = \int_0^1 e^{\frac{-SL}{Q} w_i} dG_i = \int_0^\infty e^{\frac{-SL}{Q} w_i} q_i dd_{p_i} \tag{12}$$

Here $q_i = dG_i/dd_{p_i}$ as in Chapter 3, and is not to be confused with q^*_i representing charge.

Because of the complex form of (7), as well as the possibility of complex relationships for q_i, formal evaluation of the last integral in (11) is likely to be impossible in the general case. However, for special cases some methods are available.

Whenever (8) may be taken to represent the grade efficiency, i.e., $w_i = kd_p$, and q_i is a log-normal distribution (equation (21), Chapter 2) an equation of the form of (18), Chapter 3, applies:

$$P = \frac{1}{\sqrt{2\pi}} \int_{-\infty}^{\infty} e^{-t^2/2} e^{-ae^{bt}} dt \tag{13}$$

wherein

$$t = \frac{\ln d_p/\bar{d}_{PM}}{\ln \sigma_g}$$

$a = SL\ k\bar{d}_{PM}/Q$ see equation (8)

$b = \ln \sigma_g$

Charts are available [1], [2] from which $\eta_M = 1-P$ may be read as a function of a and b, but they do not permit a very precise reading.

A unique approximation to P, developed by Kunz and Hanna [16] permits a direct calculation to be made by

$$P \simeq \frac{e^{(-A^2b^2/2-A)}}{(1+Ab^2)^{1/2}} \qquad (14)$$

where A is determined from a graph, Fig. 3, of Ab^2 vs $b\sqrt{a}$. For values of $b\sqrt{a}$ off the scale of this chart, a good calculation may be made by:

$$(Ab^2) = 2\ \ell n\ b\sqrt{a} - \ell n(2\ell n\ b\sqrt{a}) + 0.3 \qquad 10 < b\sqrt{a} < 25$$

$$(Ab^2) = \frac{2ab^2\ (1 + ab^2)}{2(1 + ab^2)^2 - a^2b^4} \qquad 0 \leq b\sqrt{a} \leq 0.5$$

The results of the calculation agree very well with values read from the charts.

Another special case, treated by Feldman [18], uses (10) as the grade efficiency function combined with a log-normal inlet particle size distribution to give

$$P = \frac{1}{\sqrt{2\pi}\ \ln\sigma_g} \int_0^\infty \exp\left[-k(aC_{d_p}) - 0.5\left(\frac{\ln d_p/\bar{d}_{PM}}{\ln\sigma_g}\right)^2\right] d\ \ln d_p \qquad (15)$$

where

$$k = \frac{\varepsilon_o E_o E}{3\mu} \frac{SL}{Q}$$

Feldman integrated this numerically for each of several size distributions typical of fly-ash from coal combustion, to obtain P as a function of k in each case. He found that ln 1/P varied linearly with ln k in a given case, so that

$$P = c^{(10^{-4}k)^m} \tag{16}$$

His data are as follows:

Furnace Type	$\bar{d}_{p_m}$-μm	σ_g	c*	m*
Pulverized Coal	12	3.8	0.5379	0.5000
Cyclone	2.08	9.45	0.8344	0.6765
Spreader-Stoker	68	3.54	0.1827	0.4055

*Values of c and m require that k be expressed in (m^4).

Where (16) is combined with (11), an expression is obtained for the parameter

$$w_e = \frac{-(10^{-k}k)^m \ln c}{SL/Q}$$

or

$$w_e = (w_e')^m (SL/Q)^{m-1} \ln (c) \tag{17}$$

where $w_e' = \varepsilon_o E_o E \times 10^{-4}/3\mu$ represents a modified rate parameter which is independent of particle size. Feldman found that (17) accounts satisfactorily for the observed behavior of w_e in a number of respects.

Example 1. Consider the same Portland cement dust described in Chapter 6, for the cyclone design example. Determine whether it could be

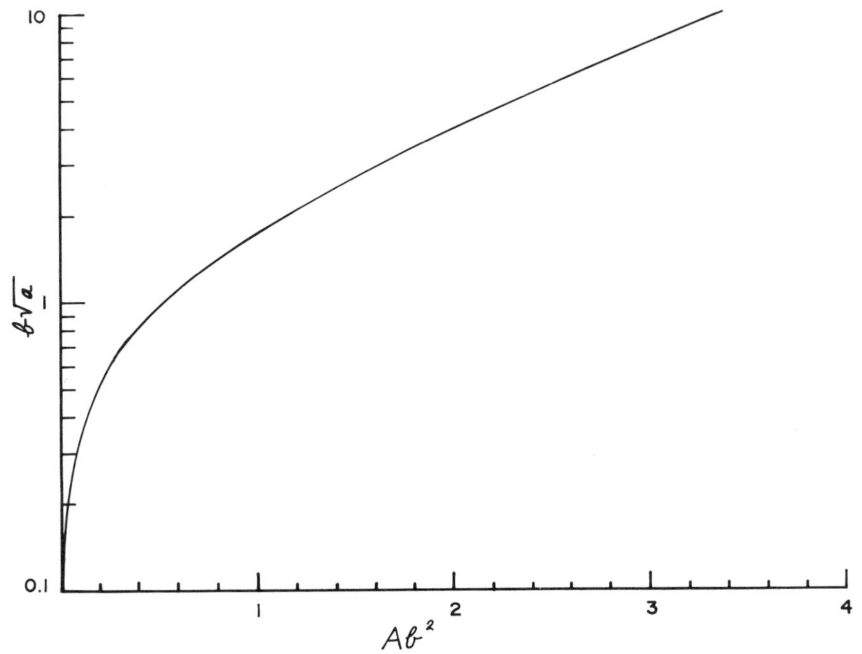

FIG. 3. Graph for use with Equation (14).

collected to meet the emission regulations by a single-stage electrostatic precipitator having 8100 ft^2 of plate surface consisting of 9 parallel plates each 15 ft. high by 30 ft. long spaced 8 in. apart. The discharge electrode potential can be as high as 32,000 V. Conditions described in Chapter 6 are: Dust: MMD = 12 μm, σ_g = 308 (log-normal distribution); Dielectric constant ε = 6.14; Flow: 222 acfs at 250°F, 1 atm.; Efficiency required: 99.78% by mass, overall.

Since this dust does not contain much submicron material (1% < 1 μm), equation (8) may be used for the grade-efficiency. Since the dust is also log-normally distributed (13) may be used to obtain the overall efficiency. Equation (8) becomes:

$$\eta_i = 1\text{-exp} - \frac{8100}{222 \times 0.0348}\left(\frac{6.14}{6.14+2}\right)\frac{8.85 \times 10^{-12}(3.15 \times 10^5)^2}{2.25 \times 10^{-5}}\ (10^{-6}d_{p_i})$$

where

$$E_o = E = \frac{32{,}000}{4/12 \times 0.3048} = 315 \text{ kv/m} = 3.15\text{x}^5 \text{ N/coul.}$$

$$\varepsilon_o = 8.85 \times 10^{-12} \text{ coul}^2/\text{Nm}^2$$

$$\mu = 2.25 \times 10^{-5} \text{ kg/ms} \qquad @250°\text{F}$$

and d_{Pi} is in microns. Thus: $\eta_i = 1 - \exp - 3.524\ d_{Pi}$. For Equation (13): $a = 3.524 \times 12 = 42.3$; $b = \ell n\ 3.08 = 1.125$. Evaluating (13) by means of (14), using Fig. 3: $b\sqrt{a} = 7.32$, $Ab = 2.91$, $A = 2.3($ and from (14): $P = 0.0018$. The overall efficiency is 99.82% which ma be taken as equal to that required, considering the uncertainties involved in the calculation. However, for this to be attained in practice there must be uniform velocity ($v_o = 222 \times 12/8 \times 15 \times 9 =$ 2.5 ft/s), uniform field potential gradient, no re-entrainment, suitable particle resistivity, ets. These factors will be investigated further in the next section.

III. CORRECTIONS TO THE MODEL

The fact that experimental values of effective migration velocity w_e (11) do not behave in all respects as predicted by the model (7) or (8) leads to consideration of various aspects of performance not incorporated in the model. These are sometimes referred to as non-Deutschian phenomena. Some of the most significant may be listed: non-uniform velocity distribution; sneakage of gas around collection zone; re-entrainment of collected particles; variation of w_e with gas velocity v_o; w_e not directly proportional to d_p; variation of w_e with width of channel between collection surfaces; decrease of w_e in full-scale unit from values obtained in pilot plant. Possible corrections to the model for these items will be discussed briefly.

A. Non-Uniform Velocity Distribution

The model assumes that the fluid velocity parallel to the collector surface is everywhere the same and equal to the volumetric rate of flow divided by the total cross-sectional area normal to the collecting electrode surface, i.e., $v_o = Q/A$. This condition is almost impossible to achieve in practice, but every effort should be made to do so. To this end it is essential to study the flow pattern in a proposed design by means of a scale model, usually 1/16 size. The effects of plate construction, baffles, shape of duct work, use of turning vanes and diffuser plate, etc. must all be studied in order to achieve the most uniform flow possible.

When it is possible to associate different values of v locally with corresponding portions of the collecting surface, a correction in (2) may be made by inserting a factor F > 1 (sometimes called a "quality factor" of the flow) with the value of v_o. Thus (2) becomes

$$P_i = \exp - \frac{SL\, w_i}{AFv_o} \tag{18}$$

F may be evaluated if the inlet velocity distribution is known in terms of a series of local values of gas velocity, called v_j, each associated with a known cross-sectional area A_j and a corresponding perimeter S_j. Then through one of these zones, the penetration P_{ij} will be, using (2):

$$P_{ij} = \exp - \frac{S_j L}{A_j} \frac{w_i}{v_j} \tag{19}$$

The total emission will be related to these values according to

$$P_i Q = \sum_j Q_j P_{ij} = \sum_j v_j \cdot A_j \cdot P_{ij}$$

or

$$P_i = \frac{1}{v_o A} \sum_j v_j \cdot A_j \cdot P_{ij} \tag{20}$$

where $A = \Sigma A_j$. If A is divided into N equal areas each A_j, this becomes

$$P_i = \frac{1}{Nv_o} \sum_{j=1}^{N} v_j \cdot P_{ij} \tag{21}$$

so that, by rearranging (18)

$$F = -\frac{SL}{Av_o} \frac{w_i}{\ell n P_i} \tag{22}$$

in which P_i is calculated from (20) or (21).

To illustrate the effect suppose that in a plug flow situation where $v_j = v_o$ the penetration would be 0.01, but that there were actually three zones of equal areas $A_j = A/3$ in which $v_1 = v_o/2$, $v_2 = v_o$, $v_3 = 2v_o$ respectively. The actual penetration would then be obtained as follows. First from (2)

$$0.01 = \exp -\frac{L}{b}\frac{w_i}{v_o} \quad \text{or} \quad \frac{L}{b}\frac{w_i}{v_o} = -\ell n\ 0.01 = 4.605$$

and from (19):

$$P_{i1} = \exp \frac{-4.605}{1/2} = 0.00010$$

$$P_{i2} = 0.01$$

$$P_{i3} = \exp \frac{-4.605}{2} = 0.1000$$

From (21)

$$P_i = \frac{1}{3v_o}\left[\frac{v_o}{2} \times 0.00010 + v_o \times 0.01 + 2v_o \times 0.1000\right] = 0.0700$$

and from (22)

$$F = -4.605/\ln 0.0700 = 1.73$$

The use of (20) or (21) with measured velocity traverses in a working precipitator may yield values of F as high as 1.5 or more in a poorly regulated flow pattern. Where care is taken, through model studies, values of F as low as 1.1 or 1.2 should be achieved.

B. Gas Sneakage and Dust Re-entrainment

Gooch and Francis [7] offer means of estimating corrections for these phenomena, means admittedly based upon over-simplified assumptions and requiring data which are not usually available. These methods would be useful however in estimating the magnitude of potential loss in performance due to these causes.

1. Gas Sneakage

This term refers to that portion of the gas flow which may by-pass the collection zone entirely. This may occur due to gas flowing through the dust collection hoppers or through the high-voltage insulation space. It is reduced or prevented by the installation of baffles which force the gas to return to the main gas passage through the collecting zone. The baffles divide the zone into a number of sections. After each section the sneakage gas is remixed into the main flow which then re-by-passes in the next section. Gooch and Francis give the following equation which may be used to correct the penetration for sneakage:

$$Ps_i = \left[S + (1-S)(P_i)^{1/N_s}\right]^{N_s} \tag{23}$$

when S = fractional amount of gas sneakage in each section; N_S = number of baffled sections; P_i = penetration of a given particle size, with no sneakage. For example, if there is 5% sneakage in each of 4 sections, an uncorrected penetration of 0.01 will be increased:

$$P_{s_i} = \left[0.05 + 0.95\ (0.01)^{1/4}\right]^4 = 0.0151$$

2. Dust Re-Entrainment

When the collection surface is rapped to dislodge the collected dust so that it may fall into the dust hopper, particles may become re-entrained in the gas stream. They may or may not be subsequently re-collected downstream. Assuming that a fixed fraction R_i of a given dust size is re-entrained and that this fraction is constant along the length of the precipitator Gooch and Francis derive the following corrected penetration:

$$P_{R_i} = \left[R_i + (1-R_i)(P_i)^{1/N_R}\right]^{N_R} \tag{24}$$

where N_R = number of stages over which re-entrainment occurs. This is of precisely the same form as the sneakage correction.

If, in addition to the sneakage loss cited above there is also a 1% re-entrainment, then

$$P_{R_i} = \left[0.01 + 0.99\ (0.01)^{1/4}\right]^4 = 0.0109$$

and the total penetration $P_S + P_{R_i} = 0.0260$, or a decrease in efficiency η_i from 99% to 97.4%. Since the precipitator is basically supposed to be a high-efficiency device obviously sneakage and re-entrainment must be minimized.

Because the calculations for sneakage and rapping re-entrainment are of the same form, they may be lumped together by use of an overall (S+R) value in place of S in (23). This is done when individual values of S and R cannot be measured, and used to back-calculate a total correction to make theoretical values agree with experimental value of P_i. Values of the lumped parameter as high as 0.10 may be found in poorly operating systems.

C. Turbulence Effects

The theoretical model gives no reason to expect that w_i should vary with gas velocity (v_o), the width of spacing between collection surfaces, or the size (scale-up) of the precipitator. It also predicts, according to (8), that w_i be directly proportional to d_{pi}. But exception to all of these expectations are readily found to occur in practice.

Cooperman has developed a theory which can account for all of these phenomena through the effects of turbulence and the consequences of particle concentration gradients in the collecting zone [9, 17]. There are three concentration gradients: in the direction of flow, "the act of collection . . . causes the concentration of particles upstream to be greater than the concentration downstream;" perpendicular to flow, due to the electric force "concentration of particles rises from a small value near the plane of the discharge electrode to much larger values in the vicinity of the collecting electrode(s);" and vertically, because of the gravity settling of particles.

If there were no turbulence, the particle concentrations would not effect the particle trajectories. They would be as described in Chapter 5 for plug flow, or for laminar flow. Because of turbulence, however, there is a mixing effect which moves particles from regions of high concentration toward regions of low concentration. This effect, called turbulent diffusion, clearly opposes the collection process, first because of the lateral gradient away from the collecting electrode, and secondly because of the downstrean (axial) gradient which causes the particles to move faster than the gas flow in this direction.

The opposition to the lateral electrostatic migration velocity (e.m.v.) w_i is taken into account by introduction of a factor f which may be regarded as the fraction of w_i which is the counterflow velocity. Then the net e.m.v. is $w_{m_i} = (1-f)w_i$.

The effect of turbulent diffusion axially is represented by the coefficient D_{td_i}, which is analogous, but not equal, to $\mathcal{D}$ used in Chapter 4 and 5 in connection with Brownian diffusion. For a parallel plate collector, a material balance over a two-dimensional (axial by lateral) element of collector volume (vertical gradient, gravity settling neglected) leads to

$$P_i = \exp\left\{ [\alpha_i - (\alpha_i^2 + 2\beta_i)^{1/2}] \frac{L}{b} \right\} \tag{25}$$

wherein

$$\alpha_i = bv_o/2\, D_{td_i}$$

$$\beta_i = bw_{m_i}/2\, D_{td_i} = b(1-f)w_i/2\, D_{td_i}.$$

α_i and β_i are dimensionless numbers representing the role of turbulence in the axial and lateral directions respectively for particles of ith size. In comparison with the Deutsch-Anderson equation (11), this is equivalent to saying that the precipitation rate parameter is replaced by

$$w_e = v\left[(\alpha^2 + 2\beta)^{1/2} - \alpha\right]$$

The relative importance of α_i and β_i may be judged better by rewriting (25) so that

$$P_i = \exp\left\{ -\alpha_i \left[(1 + \frac{2\beta_i}{\alpha_i^2})^{1/2} - 1 \right] \frac{L}{b} \right\} \tag{26}$$

Two limiting special cases may be discerned:

(a) $\alpha_i^2 >> 2\beta$, so that $\sqrt{1 + 2\beta_i/\alpha_i^2} \approx 1 + \beta_i/\alpha_i^2$

Then (25) becomes

$$P_i = \exp - \frac{\beta}{\alpha}\frac{L}{b} = \exp - \frac{(1-f)w_i}{v_o}\frac{L}{b} \tag{27}$$

This is the same as (11) with $w_e = (1-f)w_i$. It may be called the low turbulence case.

(b) $\alpha_i^2 << 2\beta_i$, so that $\sqrt{1 + 2\beta_i/\alpha_i^2} \simeq \sqrt{2\beta_i}/\alpha_i >> 1$

Then (25) becomes

$$P_i - \exp - \sqrt{2\beta_i}\, L/b = \exp - \sqrt{(1 - b)w_i L/bD_{td_i}}$$

Here $w_e = v_o \left[(1-f)w_i b/D_{td_i}\right]^{1/2}$ This may be called the high turbulence case.

Cooperman [17] points out that although no theoretical methods for calculating D_{td_i} or f are available, values for these may be obtained by back-calculating from measured performance data. In all cases f is found to be between 0 and 1, as is to be expected. Further D_{dt_i} is found to be larger for finer particles, higher gas velocities, and larger spaces, again all consistent with the concepts of turbulent diffusion. Sample calculations using D_{td} from 0.01 to 10 m^2/s and f in the range 0.5 to 0.7 yield values of w_e which are not only of the correct order of magnitude but also which conform to the observed trends.

Although no general method is available for using it, qualitatively at least, Cooperman's theory appears to be quite reasonable. It confirms the existence of a phenomenon which must not be overlooked in high turbulence precipitators.

D. Incremental Length Calculations

If the complete form of equation (7) is to be used, it is evident that conditions may change along the length of the collection zone: temperature may change, particle concentration decreases, and charging by diffusion is time-dependent. Further it is possible that values of E_o and/or E may vary. To incorporate all these effects requires that calculations be repeated for each of a series of incremental lengths, adding the collection obtained in each until the total desired collection is attained. In this way the required amount of total collection surface is calculated by finding the total L value.

Gooch and Francis [7] have outlined a computer flow diagram for carrying out such a procedure. It also incorporates the correction for gas sneakage, rapping re-entrainment, and velocity distribution, as described above. It does not incorporate the concepts of Cooperman's theory, but there is no reason they could not be added if information on f and D_{td} is available.

IV. DESIGN EXAMPLE

The author is indebted to Dr. R. Lee Byers for suggesting the following design problem and providing some of the data for it.

In Chapter 6 the use of a cyclone was considered for the collection of a Portland cement dust. The design calculations showed that a cyclone system might be useful as a primary cleaner to be followed by a high efficiency device as a secondary cleaner needed to attain the required degree of emission control. Let us now consider the design of a single stage electrostatic precipitator for this purpose.

The particle size distribution of the cement dust entering the ESP will be taken as that leaving the cyclone system, based upon 20 cyclones in parallel. This is shown as the curve for P = 20 in Fig. 7, Chapter 6. This size distribution deviates slightly from the log-normal form above $d_p \simeq 12$ μm, but the dust may be well represented by using an MMD = 3.7 μm and $\sigma_g = 2.22$.

Other conditions will be as follows: Gas flow rate, 13,300 acfm; Gas temperature 240 F; Gas pressure, 1 atm.; Gas viscosity, $\mu = 2.25 \times 10^{-4}$ poises; Dust properties: $\varepsilon = 6.14$, Inlet c_o, 1.24 gr/acf, Rate 141.5 lbm/hr.; Outlet c_o, 0.0132 gr/acf, Rate 1.5 lbm/hr.; Density, $\rho_p = 1.5$ gm/cm^3; Efficiency required: $\eta_M = 98.94\%$. For data relating to similar installations consult reference [4], Tables 18.10, 18.12 and Fig. 18.10. These suggest that it would be appropriate to assume the following additional conditions. Field strength $E_o = E = 8$ kv/in. Average velocity $v_o = 2$-8 ft/s.

Since only a few percent of the dust is finer than 1 micron, the calculations will be carried out using equation (8) as the basis for the grade-efficiency. The set-up is similar to that for the previous example, page 18, except that (L) is the unknown to be determined. Using the numbers presented there:

$$P_i = \exp - Md_{pi} = \exp -4.351 \times 10^{-4} \text{ (SL) } d_{pi}$$

with $M = 4.351 \times 10^{-4}$(SL) μm^{-1}, (SL) in ft^2 and d_{pi} in μm.

The required performance to be attained is that according to equation (12), Chapter 3

$$\eta_M = \int_0^1 \eta_i dG_i = 0.9894$$

or according to equation (17), Chapter 3

$$P = \sum_i \Delta g_i \cdot P_i = \Sigma \Delta g_i \cdot \exp - Md_{pi} = 0.0106$$

There are two methods available for solving for M. The first is by using the Kunz and Hanna approximation to the integral for η_M, given as equation (14) above. The second is to solve the series for P by the methods of equations (24) and (25) of Chapter 3. For the first, the inlet dust must be represented by a log-normal size distribution. For the second, the actual distribution may be used. Both methods will be illustrated here for purposes of comparison.

For the Kunz and Hanna method, equation (14): MMD$\equiv \bar{d}_{Pm} = 3.7\ \mu m$, $\sigma_g = 2.22$; $k \equiv M = 4.351 \times 10^{-4}$ (SL) μm^{-1}, (SL) = ft^2; $a = k\bar{d}_{Pm}$ $= 1.573 \times 10^{-3}$ (SL); $b = \ln \sigma_g = 0.7975$ and $P = 0.0106$. Then

$$P = \frac{e^{-0.3180A^2-A}}{(1+0.6360A)^{1/2}} = 0.0106$$

must be solved for A, This is best done by an iterative procedure, which may easily be programmed on a pocket calculator in the form:

$$A = \left[\frac{(-\ln 0.0106)(1+0.6360A)^{1/2}-A}{0.3180}\right]^{1/2}$$

It yields A = 2.3438. Then $Ab^2 = 1.4907$ and $2.60 = a^{1/2}b$ from Fig. 3. Finally $a = (2.60/0.7975)^2 = 10.62$ (M = 2.870) and SL = $10.62/1.573 \times 10^{-3} = 6760$ ft^2.

For the series procedure, it is necessary to choose an interval for Δg_i, then tabulate the corresponding values of d_{Pi}. Two intervals will be used: 0.05 and 0.10. The following values are read from the plot for P = 20, Fig. 9 Chapter 6.

i	$\Delta g=0.05$		$\Delta g=0.10$	
	Δd_P	d_{Pi}	Δd_P	d_{Pi}
1	0-0.95	0.475	0-1.28	0.64
2	0.95-1.28	1.115	1.28-1.83	1.56
3	1.28-1.58	1.43	1.83-2.38	2.11
4	1.58-1.83	1.70	2.38-2.95	2.67
5	1.83-2.10	1.97	2.95-3.65	3.30
	etc.		etc.	

A maximum of five terms in the series should be ample in order to solve for M, in each of the following equations:

$$\Delta g = 0.05: \quad \frac{0.0106}{0.05} = e^{-0.475M} + e^{-1.11M} + e^{-1.43M} + \ldots\ldots$$

$$\Delta g = 1.10: \quad \frac{0.0106}{0.10} = e^{-0.64M} + e^{-.156M} + e^{-2.11M} + \quad \ldots\ldots$$

Again, these solutions may readily be carried out by programming an iterative procedure on a pocket calculator in the form (for example):

$$M = -\frac{1}{0.64}\,\ell n\left[0.1060 - e^{-1.56M} - e^{-2.11M} \quad \ldots\ldots\ldots\right]$$

The results are tabulated

Terms Used	Δg=0.05 M	Δg=0.10 M
1	3.266	3.507
2	3.481	3.564
3	3.538	3.571
4	3.559	3.572
5	3.565	3.573

Both methods appear to converge to the same value if a sufficient number of terms is used. For a quick approximation, using one term of the Δg=0.10 series will be the best.

The required plate area SL = 3.573/4.351 x 10^{-4} = 8210 ft^2 by this method. Note that this result is based entirely upon the size distribution data in the fine end of the range. It uses only the lower 20% (Or 40%) of the mass, according to the Δg_i selected. However, this is the critical material to be collected, in order to meet an emission limitation. The coarser material will assuredly be collected with 100% efficiency.

For the reason cited, and because it is the larger value, the design should be based upon 8200 ft^2 of collection surface rather than the value of 6760 ft^2 obtained by the Kunz and Hanna method. The latter involves some mathematical approximation and also the reading of a chart, both points introducing an unknown degree of error.

The value of w_e, the uncorrected "effective migration velocity" or "precipitation rate parameter" will be, from (11)

$$w_e = -Q\ell n\ P/SL = -222\ell n\ 0.0106/8210 = 0.123\ \text{ft/s.}$$

Allowance should be made for uneven velocity distribution, for gas sneakage, and for rapping re-entrainment. According to the methods and examples given above we may take (for all particle sizes) F = 1.2 in (18) and (S+R) = 0.05 in (23) or (24), assuming N = 5. For a corrected P_{S+R} = 0.0106, the uncorrected value from (23) must be

$$0.0106 = \left[0.05 + (0.95)(P)^{1/5}\right]^5$$

$$P = \left[\frac{(0.0106)^{1/5} - 0.05}{0.95}\right]^5 = 0.00706$$

The corrected precipitation rate parameter will then be:

$$w_{e_{corr}} = 0.123 \times \frac{\ell n\ 0.0106}{\ell n\ 0.00706} \times \frac{1}{1.2} = 0.0941\ \text{ft/s.}$$

and the corrected surface (SL) will be

$$(SL) = -\ 222\ \ell n\ 0.0106/0.0941 = 10{,}730\ \text{ft}^2$$

The remainder of the design involves specifying plate dimensions and spacing. Assuming v_o = 2.5 ft/s, plate spacing of 8", and plate height of 15 ft, would require

$$\frac{13{,}300}{60} \times \frac{12}{2.5 \times 15 \times 8} = 8.87\ \text{spaces}$$

Taking 9 spaces will require 10 plates, each 15 ft high by

$$\frac{10,730}{15 \times 10 \times 2} = 35.8 \text{ ft. long}$$

and the actual value of

$$v_o = \frac{222}{19 \times 8/12 \times 15} = 2.47 \text{ ft/s.}$$

The assumption of N = 5 sections seem reasonable for a total plate length of nearly 36 ft. This would make the length in each section just over 7 ft (7.2 ft).

In comparing these results with the previous example it is noted that a much greater collection surface is required even though the allowable penetration is greater. This is due in part to the corrections for poor velocity distribution, sneakage and re-entrainment, and in part because the particle size distribution is finer in this case.

REFERENCES

1. H. White, *Journ. Air Poll. Contr. Assoc.*, 27: 15, 114, 206, 308 (1977).

2. M. Robinson in *Air Pollution Control-Part I*, W. Strauss, ed. Wiley-Interscience, New York, 1971.

3. S.J. Oglesby et al, *A Manual of Electrostatic Precipitator Technology-Part I Fundamentals*, U.S. Dept of Commerce NTIS, PB 196, 380, Aug. 25, 1970.

4. Ibid-*Part II Application Areas* PB 196, 381, Aug. 25, 1970.

5. Ibid-*Selected Bibliography* PB 196, 379, Aug. 25, 1970.

6. G.B. Nichols and S.J. Oglesby-Ibid-Part I, pg. 188.

7. J.P. Gooch and N.L. Francis, *Journ. Air Poll. Cont. Assoc.*, 25: 108 (1975).

8. J.P. Reynolds, L. Theodore, and J. Marino, *Journ. Air Poll. Cont. Assoc.*, 25: 610 (1975).

9. P. Coopermann, *Atmos. Environ.*, 5: 541 (1971).

10. M. Robinson, *Atmos. Environ.*, 1: 193 (1967).

11. J.H. Abbott and D.C. Drehmel, *Chem. Eng. Prog.*, 72: 47, (1976).

12. M. Crawford, *Air Pollution Control Theory-Chap. 8*, McGraw-Hill, New York, 1976.

13. G. Leutert and B. Bohlen, *Staub*, 32: 27 (1972).

14. J.P. Gooch and J.R. McDonald, *A.I.Ch.E. Symp. Ser.* 73: 146 (1977).

15. W.B. Smith and J.R. McDonald, *Journ. Air. Poll. Cont. Assoc.*, 25: 168 (1975).

16. R.G. Kunz and O.T. Hanna, *Ind. Eng. Chem. Process Des. Develop.*, 11: 623 (1972).

17. P. Cooperman, Paper No. 76-42.2, 69th Annual Meeting *Air Poll. Cont. Assoc.*, 1976.

18. P.L. Feldman, Paper 75-02.3, *Air Poll. Control Assoc.*, Boston (1975).

19. R. Cochet, *Compt. Rendues* 243: 243 (1956).

20. R. Cochet, *Colloq. Intern. Centre Natl. Rech. Sci.*, (Paris) 102: 331 (1961).

CHAPTER 8

FILTRATION

I. GENERAL CONCEPTS

Filtration may be defined as a dry method of particulate collection in which an array of many individual targets is assembled into a porous structure through which the aerosol-laden gas is passed. The aerosol particles impinge on the targets due to any of the modes of aerodynamic capture and are held there. As collection proceeds, particles impinge upon previously collected particles and a deposit is built up which in turn may itself become the principal collecting medium. The collected dust must eventually be removed somehow or it will hinder the gas flow to an intolerable degree. Filtration is therefore a repeated cyclical process of collecting and cleaning.

There are basically three types of arrays employed: (a) individual fibers loosely packed into a pad or mat; (b) granular material in either a fixed, moving, or fluidized bed; (c) fibers, woven or felted into a fabric, which is used in the form of a bag. All of them are capable of achieveing very high collection efficiency with reasonable pressure drop up to the time of cleaning. Filtration is therefore used to collect very fine particulate under dry conditions. For an extensive description of types of devices and of theoretical principles two principal references may be cited: the book *Air Filtration* by C.N. Davies [1] and the state-of-the-art study *Handbook of Fabric Filter Technology* by Billings and Wilder[2].

The principal limitation to its application lies in the nature of the material which may be used for the target fibers or granules. For fibrous filters or fabric filters, a wide variety of natural and synthetic fibers are available. It is necessary to select one which will have a suitably long life at reasonable cost under the conditions prevailing in a given problem. For example, a polymer (e.g., nylon, Teflon) or inorganic (e.g., fiberglass) fabric may be specified to meet the conditions of higher temperatures or contact with corrosive gas. Slag-wool or metallic mesh may be used in fibrous mats. Granular bed filters may be made of inert inorganic granules (sand, gravel, ceramics) capable of operation at very high temperatures.

The method of cleaning is also very important. Fabric used in the form of a bag or tube is cleaned mechanically by shaking, or blowing gas in the reverse direction. The material must have sufficient mechanical strength to withstand the repeated flexing involved. Fibrous pads are either discarded and replaced periodically or may be washed and/or dumped periodically, then repacked. Moving beds or fluidized beds have the possibility of being cleaned mechanically and continuously.

A understanding of the initial collection process must be based upon the principles of aerodynamic capture discussed in Chapter 5. The principal mechanisms involved are inertial impaction, Brownian diffusion, and in some cases electrostatic charging. However, after the initial collecting period particles are captured upon previously collected particles in what may be a random and erratic fachion. This process is not well understood, so that aerodynamic capture theory is primarily useful only in predicting the performance of a clean filter.

Furthermore, in the case of fabric filters some new phenomena come into play which are not covered by the concepts of aerodynamic capture. Some particles may pass straight through without being captured at all. Collected particles are observed to seep through the layer of dust and fabric under the continued flow of gas. "Pinholes" may form which permit little plugs of particles to loosen and pass through. These are referred to as "fault processes" and they may be the principal deterrent to otherwise attaining 100% collection by use of a fabric.

Finally, the target elements themselves in actual practice are not like the idealized targets (cylindrical fibers, spherical granules) treated in capture theory. They are irregular in shape and random in orientation in the array.

For all of these reasons, the design of filtration collectors is very much more of an art based upon prior experience than is the design of other kinds of collectors. It is worthwhile to begin the design by calculating the magnitude of the several dimensionless groups involved in aerodynamic capture, as listed in Table 4, Chapter 5. This will serve to identify the important mechanisms, and to understand at least qualitatively the performance trends which may be expected, as tabulated in Table 4, Chapter 5. For fibrous and granular filters the theory of arrays as given in Chapter 5 may be used to estimate collection efficiency. However, this may not be very applicable to fabric filters.

The pressure drop across a filter is also an important factor in design. It tends to increase in proportion to the amount of dust collected. The maximum pressure drop, just before cleaning or replacement, must be estimated in order to determine the power required for the operation. Usually this is the limiting factor which determines when cleaning must begin. Some theories and empirical relationships are available for estimating ΔP.

II. FIBROUS FILTERS

A. Collection Efficiency

1. Initial Capture

The overall efficiency of a clean filter may be estimated by use of equations (74) or (75), Chapter 5, based upon an array of fibres:

$$P_i = \exp - \eta_{T_i} \frac{(1-\varepsilon)}{\varepsilon} \frac{4L}{\pi D} = \exp - \eta_{T_i} \frac{D\ell'}{\varepsilon} = \exp - S\eta_{T_i} \tag{1}$$

Filters of this type are characteristically made up of fibres ranging from approximately 0.5 μm to 50 μm diameter, with porosity values usually greater than 0.95 and thickness ranging from a millimeter to several centimeters. Some typical actual values, together with the calculated solidarity factors (S) and calculated values of the total effective fibre length per unit cross-section of filter (ℓ') are:

D-μm	ε	L-cm	S	ℓ'-μm^{-1}
50	0.995	3.0	3.8	0.0756
3	0.997	1.5	19.2	6.38
11.	0.951	0.71	42.3	3.66
1.24	0.925	0.05	41.6	31.0
0.50	0.970	0.13	102.	197.9

The values of D quoted are either based upon a uniform fibre diameter or are an appropriate effective mean fibre diameter.

Values of the single fibre efficiency (η_{T_i}) in the array, may be obtained by the theoretical principles outlined in Chapter 5 using the Kuwabara flow field. The most extensive study of this kind is that of Yeh and Liu [3]. It takes into account inertial impaction, direct interception, and Brownian diffusion operating simultaneously. Electrostatic and gravity effects are assumed to be negligible. The Kuwabara model is modified to allow for consideration of slip-flow at low Knudsen numbers based upon the fibre diameter: $Kn_f = 2\lambda/D$.

The velocity field, as determined by Yeh [4], reduces to that given by equations (59), Chapter 5, as $Kn_f \to 0$. The value of the Kuwabara number Ku, given by (60), Chapter 5, is modified to

$$\gamma = \frac{1}{2} \ell nc - \frac{3}{4(1+Kn_f)} + \frac{c}{1+Kn_f} + \frac{Kn_f(2c-1)^2}{4(1+Kn_f)} - \frac{c^2}{4} \tag{2}$$

and $\gamma \to Ku$ as $Kn_f \to 0$.

Unfortunately the only way to obtain η_{Ti} by Yeh and Liu's method is to do the numerical solution of the differential equations for particle trajectories. They present results of this in graphical form and compare them with the results of the method outlined by Davies and given by equation (63), (64), and (65) in Chapter 5. Within the limitations of $R<0.1$, $Pe>100$, stated there the agreement is good at low Kn_f. As R increases above 0.1, or as Kn_f increases above 0, there is considerable disagreement and Yeh and Liu's charts would have to be used. Their theoretical results are well substantiated by their experimental measurements [5], for the case of a uniform value of D.

The effect of a mixture of fibre sizes in a filter has been considered by Clarenburg and Van der Wal [6]. They found that provided a mean effective fibre diameter $\bar{D}_E$ was properly defined for the mixture, the porosity of the filter affected the penetration according to

$$-\ell n\ P = \Gamma \frac{L}{D_E^2} \frac{(1-\varepsilon)^2}{\varepsilon} \tag{3}$$

where Γ was independent of the composition of the filter. The definition of D_E is given by

$$\log \overline{D_E^2} = \sum_{j=1}^{n} x_j \log \overline{D_j^2} \tag{4}$$

where j designates fibres of a certain diameter present in weight fraction x_j (corrected for the shadow effect of one fibre over another). The porosity function in (3) is confirmed experimentally [6]. To reconcile (3) and (1) would require knowledge of how η_{Ti} is related to Γ, which must be a function of particle size. This is not available, but presumably the net effect of ε upon η_T is such as to introduce the additional $(1-\varepsilon)$ factor in (3). For filters of a given composition, i.e., given "mix" of fibres, the effect of varying degrees of compaction upon collection efficiency can be determined by (3).

The calculation of η_{T_i} by the procedure outlined above is not always as elaborate as that may seem. If the various dimensionless groups are calculated first, (see Table 4, Chapter 5) usually it will be found quickly that one mechanism predominates and subsequent calculations need only be done for that one.

However, all of the above material pertains only to the initial performance of a clean filter. As particles accumulate, further capture takes place on particles already collected. Some recent theoretical studies have attempted to take this into account but no practical scheme of calculation has yet become available. In general it is observed that collection efficiency tends to increase, sometimes exponentially, as collection proceeds. Design based upon the behavior of the clean filter will therefore be conservative.

2. Adhesion and Retention

It is well recognized in all the theory of aerodynamic capture, that just because a particle collides with a target it does not necessarily adhere to the target. Adherence may be more likely if collision is due to Brownian diffusion or electrostatic forces, than if it is due to inertial impaction or direct interception. Adherence is certainly greater on fibres which have been coated with a liquid than upon dry fibres. Retention is more difficult at larger particle sizes because they are more likely to be subject to "bounce-off" or the subsequent "blow-off" (re-entrainment by drag). These effects have been described in terms of a "retention efficiency" η_R, acting upon the initial aerodynamic capture or "collision efficiency" η_c, such that the effective filter efficiency is given by

$$\eta_s = \eta_c \times \eta_R \tag{5}$$

A series of experiments by Freshwater and Stenhouse [7] have elucidated some of the factors affecting η_R. These experiments were done mostly on relatively coarse ($d_p > 1\mu m$) particles. They investigated the effects of (a) relative humidity of the air (% R.H.); (b) particle size (d_p); (c) air velocity (v_o) up to 600 cm/s.; (d) dust loading. The forces of adhesion increased with % R.H. and became relatively constant between 35% and 55%. According to inertial impaction theory, the single target efficiency should increase both with d_p or v_o, as they enter into Stk. But it was found that as either d_p or v_o was increased, η_R (and η_s) began to decrease above certain values as the bounce and/or drag effects increased. When fibres were oiled these effects disappeared. The same phenomena were noted with respect to dust loading. At higher loading, evidently agglomerates of captured particles build up and are more subject to removal by air drag.

The operation of filters at high velocities is very desirable because it reduces the cross-sectional area needed to handle a given volume of gas. It also may make inertial impaction a significant collection mechanism even on sub-micron particles. Consequently an entirely different line of experiments was carried out by First and Hinds [8] using sub-micron particles, both solid and liquid, and very high velocities (up to 1524 cm/s). The filter studied was a fiberglass mat composed of fibres having a median diameter of 4.1 μm, and packed to a porosity of either 99.9% (1.9 cm thick) or of 97.7% (0.40 cm thick).

The behavior of the liquid and the solid particles was essentially different as shown by the data tabulated below. These efficiencies were determined by light-scattering measurements.

Efficiency η_s

- at v_o (cm/s) -

d_p-μm	250	500	750	1000	1250	1500
		Liquid (DOP) aerosol				
0.80	0.14	0.85	0.96	0.98	0.99	0.99+
		Solid (PSL) aerosol				
0.36	0.46	0.79	0.87 (**)	0.86	0.74	0.73
0.56	0.37	0.75	0.85	0.78	0.68	0.65
0.80	0.43	0.88	0.72	0.55	0.61	0.70
1.10	0.72(*)	0.63	0.55	0.56	0.72	0.83

(*) Maximum efficiency 0.75 at v_o = 300 cm/s

(**) Maximum efficiency 0.88 at v_o = 900 cm/s

The liquid particles are captured in almost quantitative accord with capture theory for impaction and interception, that is η_R in (5) is essentially equal to 1. The solid particles exhibit a maximum efficiency at a velocity which is higher the smaller the particle. A "blow-off" re-entrainment appears to set in with η_R increasing as v_o increases. The presence of another reversal of trend at the highest velocities for the two largest particles sizes is unexplained.

No quantitave methods have been developed by which values of η_R may be correlated or predicted.

B. Pressure Drop

The concept of "number of inlet velocity heads" N_H, discussed in Chapter 3, is not used in relation to filters because the flow is almost always under conditions of laminar motion, i.e., $\Delta P \propto v_o$ (although not necessarily at the high velocities studied by First and Hinds [8]). Provided that Kn_f is low enough that slip past the fibres is not significant, at $Re_D < 1$, Darcy's law may be taken as the basic relation for ΔP, in the form

$$- \Delta P \cdot D^2 / L v_o \mu = f(\varepsilon) \tag{6}$$

The quantity $v_o L/\Delta P$ is called the "permeability" of the filter and is a unique function of the fibre size and bed porosity for geometrically similar filters in the same fluid.

Davies [9] found the form of $f(\varepsilon)$ in (6) to be

$$f(1-\varepsilon) = 64[(1-\varepsilon)^{1.5}\ 1 + 56(1-\varepsilon)^3] \tag{7}$$

by experiments with a variety of filter pads. This was also confirmed by Werner and Clarenburg [10]. Values of ε ranging from 0.70 to 0.994 were covered. For mixtures of fibres a mean square diameter D_E^2 must be used similar to that in (4) above. From either the Happel or the Kuwabara flow field theory (see Chapter 5), a theoretical expression for $f(\varepsilon)$ may be derived. These expressions give results which lie on either side of (7) depending upon whether the fibres are assumed to lie parallel or transverse to the direction of flow.

At higher flow rates (higher Re_D), and especially for relatively thick pads, Darcy's law is no longer followed both because of turbulence and also because the pads tend to be compressed. A comprehensive study by Wright, Stansny and Lapple [11] indicates that the effect of Re_D may be taken into account by a modified form of (6):

$$-\Delta P \cdot D^2 / L v_o \mu = 8(1 - \varepsilon) C_f \tag{8}$$

where C_f = flow coefficient, obtained from a chart, Fig. 1, as a function of Re_D and porosity ε.

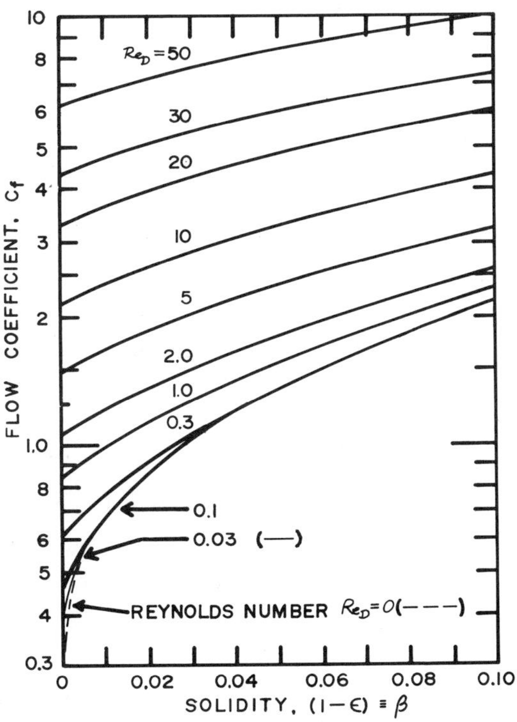

FIG. 1. Flow coefficient for pressure drop in fibrous beds [11]. See equation (8).

A formula proposed by Kimura and Iinoya [27] for glass and steel fibrous filters, correlating a lot of experimental data, may be written in the form of (8), giving

$$-\Delta P \cdot D^2/Lv_o\mu = (0.6Re_D + 4.7\sqrt{Re_D} + 11)\ 2(1-\varepsilon)/\pi \tag{8a}$$

This would correspond to the flow coefficient C_f, given by

$$C_f = (0.6\ Re_D + 4.7\sqrt{Re_D} + 11)/4\pi \tag{8b}$$

Values of C_f so calculated agree well with those read from Fig. 1, at porosities $\varepsilon \approx 0.97$ for $Re_D \leqq 5$. Above $Re_D = 5$, the calculated values correspond to larger ε's, and at $Re_D = 50$ the $C_f = 5.9$ is off the chart. Of course (8b) does not indicate an influence of ε upon C_f such as is given by the abscissa of Fig. 1.

Example. A filter is to made up of fibres having an effective diameter of 10 μm and a density of 2.34 gm/cm^3, packed with a porosity of 97.0%. It is proposed to use it to collect dust having a mass median equivalent diameter of 0.5 μm with an initial efficiency of at least 90% by weight. The gas flow is to be 500 acfm at 20 C, 1 atm. Determine the cross-sectional area and thickness of the filter required, and estimate the pressure drop across it.

For the sake of simplicity we begin this design by using only the MMD particle size, laying aside the question of particle size distribution until later.

The cross-sectional area, thickness, and pressure drop of the filter are all interrelated through the lineal velocity. If a value of v_o is specified each of these items may be determined. However, there is certainly a range of possible values for v_o, and the optimum value must be sought. We illustrate the process by selecting two values: 40 ft/min. and 100 ft/min.

For each v_o selected, all of the dimensionless parameters (Re_D, Stk, Pe^{-1}, G·Stk, R, C, Ku, J) must first be calculated. Then the single target capture efficiency for each important mechanism must be found and these values combined into the overall target efficiency for the array. The procedure is as outlined in Chapter 5 and illustrated by example 7 there. To do this we need values for C = 1.33, and $\mathcal{D}$ = 6.26 x 10^{-7} cm^2/s, for the d_p = 0.5 μm.

The results of these calculations are as follows:

v_o	40 fpm	100 fpm	Eqn.
Re_D	0.135	0.338	
Stk	0.048	0.120	
R	0.05	0.05	
Pe^{-1}	3.1 x 10^{-5}	1.2 x 10^{-5}	
G·Stk	1.15 x 10^{-4}	4.6 x 10^{-5}	Chap. 5: (54)
c	0.03	0.03	
Ku	1.033	1.033	Chap. 5: (60)
J	0.058	0.058	Table 6, Chap. 5
η_{II}	0.00269	0.00674	Chap. 5: (64)
η_{DI}	0.00234	0.00234	Chap. 5: (63)
η_{BD}	0.00375	0.00208	Chap. 5: (65)
η_G	0.00011	0.00004	Chap. 5: (54)
η_T	0.0089	0.0112	Chap. 5: (62)

Using these results, the value of L is obtained from Chap. 5 (74) and ΔP from (7) above. The area is found from setting v_o = Q/A.

v_o	40 fpm	100 fpm
S	259	206
L	6.58 cm.	5.23 cm.
ΔP	3.2 "H_2O	8.0 "H_2O
A	12.5 ft^2	5 ft^2

As the velocity increases the cross-sectional area and thickness of the filter decrease and the pressure drop increases. The optimum value of v_o will have to be determined by an economic balance over initial cost (A and L) against operating cost (power, ΔP).

A more complete design taking into account the particle size distribution would follow the procedure illustrated by the design example for electrostatic precipitators, Chapter 7. Here for each value of v_o a set of parameters corresponding to several size intervals would be calculated, and a value of η_{T_i} for each interval determined. Then S would be found by solving an equation of the form:

$$P = \Delta g_i \sum_i e^{-S\eta_{T_i}}$$

Since η_{T_i} would increase rapidly with the value of d_{p_i}, the higher terms of this series would drop off rapidly. A solution could be obtained using only the first few terms as was done in the ESP example.

III. GRANULAR BED FILTERS

The use of a granular bed (composed for example of sand or pebbles) as a filter is an old idea which has recently received renewed attention because of the possibility of using it at high temperatures and high pressures. Such conditions will be encountered in the development of gas cleaning from "advanced" energy processes such as coal gasification and fluidized bed combustion.

Granular beds may be classified as fixed, intermittently moving, continuously moving, or fluidized. The type of operation is primarily related to the method of cleaning. Cleaning requires removing the deposited dust from the granules either mechanically (shaking or moving the bed) or by back-flow of air. Fluidized beds may be continually renewed by fresh granules.

The basic model may be taken to be an array of spherical targets as presented in Chapter 5. The porosity of granular arrays is much lower than that of the fibrous filters discussed above. Fluidized or moving beds will have a higher porosity than fixed beds which usually will be of the order of $\varepsilon = 0.40 - 0.50$.

A. Fixed Beds

This term will be taken to include all beds in which the targets are in contact with each other and not moving relative to each other during the collection process. The porosity of fixed beds is much lower than that of fibrous filter, in the range of 0.35 to 0.60 usually. All of the following material relates to the behavior of a clean bed.

1. Collection Efficiency

Because of the increasing interest in granular beds the U.S.E.P.A. has commissioned a thorough review and study of them. A review report was presented by Yung, Patterson, Calvert, and Drehmel [12]. After studying several models which have been proposed, as well as published experimental data and some new data of their own, they concluded that: (1) For values of $v_o < 10$ cm/s; the array behavior given by equation (76), Chapter 5, is valid; the total single target efficiency may be predicted by Paretsky's model, equations (67), (68), (69), (70), Chapter 5, involving II, DI, BD, and G mechanisms. (2) For values of $v_o > 10$ cm/s: inertial impaction is the primary mechanism, but none of the available models is satisfactory. There is a model for II only, proposed by Jackson and Calvert [13], which may be adjusted to give a good fit. Starting with equation (76), Chapter 5, η_T is replaced by (2 Stk), and $3(1-\varepsilon)/2\varepsilon$ is replaced by the empirical constant C.

2. Pressure Drop

Ergun's general equation [14] for flow through packed beds is found to be applicable to the clean granular filter beds tested.

$$-\Delta P = f \frac{v_o^2 \rho_f L}{D} (1-\varepsilon)/\varepsilon^3 \qquad (9)$$

where

$$f = 150/Re'_D + 1.75 \qquad (10)$$

and

$$Re'_D = Dv_o\rho_f/\mu_f(1-\varepsilon) \tag{11}$$

It may be noted that at low values of Re'_D, the second term in (10) becomes negligible and (9) reduces to a form of Darcy's law like (6) in which $f(\varepsilon) = 150(1-\varepsilon)^2/\varepsilon^3$. At higher values of Re'_D, (9) may be expressed in terms of inlet velocity heads like equation (48), Chapter 3, in which

$$N_H = 2g_c \frac{fL}{D} (1-\varepsilon)/\varepsilon^3 \tag{12}$$

At sufficiently large values of Re'_D, $f \to 1.75$, and N_H becomes a function only of the bed target geometry (L, D, ε).

B. Fluidized Beds

A fluidized granular bed may be contrasted with a fixed one by noting that the particles are moving about and are not in contact with each other, the porosity is much higher, and gas passes through in discrete bubbles. The bubbles provide a means for particulate to pass through the bed without contacting the collecting target particles of which the bed is composed. When a fixed bed is fluidized by gradually increasing the gas velocity, the overall collection efficiency is found to decrease markedly at and above the minimum v_o required for fluidization (v_{mf}). This is attributed to the onset of the bubbling process [15], [16], [17].

Although fluidized beds can be continuously renewed without interrupting the filtering process, their use would not be attractive except for the possibility of multistage operation. Two to five stages may be superimposed vertically in a column so as to give the effect of filters in series. McCarthy, et.al. [16] have shown that this arrangement will maintain high collection efficiencies on fine particles ($d_p \simeq 0.1$ μm) at velocities up to 3 v_{mf}. On any given particle size the overall efficiency of n stages will be related to that of a single stage (η_1) by equation (77), Chapter 5. Thus

η_1	Overall Efficiency			
	n = 2	n = 3	n = 4	n = 5
0.60	0.840	0.936	0.974	0.990
0.80	0.960	0.992	-	-
0.90	0.990	0.999	-	-
0.95	0.997	-	-	-

These calculations assume, of course, that each stage operates at the same efficiency and that there is complete mixing of uncollected particles between stages.

Some indication of the collection mechanism operating in a fluidized bed may be obtained by sample calculations assuming spherical targets. The target diameters are typically in the range of a few hundred microns. For sub-micron particulate and velocities in the neighborhood of v_{mf} (a few cm/s) these calculations show that II is insignificant and the collection is primarily due to DI for the coarser particles and BD for the finer ones. If the bed particles become charged, ES forces may increase collection enormously [17].

The array models presented in Chapter 5 are not applicable to fluidized beds because of the random motion of the collecting targets, as well as to the presence of the gas bubbles. No theoretical model has been presented.

Pressure drop across a bubbling bed may easily be estimated. It is equal to the total weight of bed particles divided by the cross-sectional area of the bed. The total frictional drag on all the suspended particles is just equal to the force required to suspend them: their weight.

IV. FABRIC FILTERS

Fabric filters differ from fibrous filters in that the fibers are woven into a fabric which has a much lower porosity than the loosely packed fibrous pads or mats. After a new fabric has gone through several cycles of use and cleaning, it retains a residual cake of dust which becomes the filter medium.

The cloth fabric is usually made into cylindrical tubes, of the order of a foot in diameter by 20 or more feet long. A number of these tubes are suspended vertically and operate in parallel in a unit called a "bag-house." They are cleaned either by mechanical shaking or by a reverse flow of air. Most of the collected dust drops off into a hopper below although some of it is resuspended in the air and refiltered. The fabric must have good strength to withstand the repeated flexing which occurs during cleaning.

The use of cloth fabric also brings in another important difference, namely that the temperature of operation is limited by the properties of the fabric. In many cases this requires that the gas/dust stream be precooled before entering a fabric filter. The cooling system must become an integral part of the design.

The nature of the capture process may be viewed as analogous to that in fibre filters, but this is probably an over-simplified approach. The situation is more complex than that modelled by the array of cylindrical targets as explained in Chapter 5. Billings, [18] has viewed it on three levels of scale:

- molecular, dealing with molecular surface interactions of individual particles and fibres; scale $< 10^{-5}$ cm
- microscopic, dealing with local flow through localized deposit of several particles and neighboring fibres; scale $\sim 10^{-2}$ cm.
- macroscopic, dealing with behavior of fabric, bag, and dust deposit as a whole; scale 10^{0} cm.

At each level he has enumerated many factors which affect the performance. First and Leith [19] have found that a significant part of the penetration is due to "fault processes" which permit particles to continue to be emitted from the deposited cake although no fresh particles are entering the filter. In spite of these considerations a system of modelling based upon the target array theory continues to be used and will be outlined below.

However, the interaction of a particular dust with a particular fabric, under a given set of conditions seems to be highly specific. In consequence filter design must be based to a very large extent upon performance experience, more so perhaps than for any of the other collectors. This may be organized in a systematic way by the system analysis shown in Fig. 2. This will also be discussed below, along with more details of the First and Leith findings.

A. Fibre Array Model

Early approaches to modelling the fabric filter followed precisely the same scheme as that given above for fibrous filters. The basic mechanisms of aerodynamic capture were assumed to account for the collision of particles with fibres, although it was recognized that fibres making up a fabric are curved and intertwined, and are represented only poorly by an array of straight, parallel cylinders. These mechanisms no doubt do indeed account for the initial capture in a clean fabric. But they do not take into account the subsequent behavior of the composite residual-dust/fabric-medium as a collector, nor do they account for the "fault processes."

A representative recent model of this kind is that of Fraser and Foley [20]. They developed descriptions for: (1) pre-conditioning to cool the gas; (2) fallout of dust into hopper prior to reaching fabric; (3) build-up of pressure across the bags; (4) collection efficiency of the bags; (5) the effect of cleaning on the residual dust and drag. The pre-conditioning involves standard engineering calculations related to the heat exchange involved in the cooling process, which will be illustrated in the design example below. The fallout of dust pertains to the gravity settling of larger particles against the rising current of gas entering the bags. It is based upon the calculation of terminal settling velocities as presented in Chapter 4. The build-up of pressure drop across a bag is not treated by an equation such as (6) or (8) above. Rather it is based upon a semi-empirical approach which is also used in the system design approach discussed below.

Collection efficiency in the Fraser and Foley model is based upon equation (74) Chapter 5 for the array, using modified Davies equation (57) Chapter 5 for the single target efficiency. This equation represents the combined effect of II, DI, and BD for an $Re_D = 0.2$, around an isolated cylinder. To allow for the effect of neighboring fibres a correction proposed by Chen [21] is used:

$$\eta_{T_i} = \eta_{s_i} (1 + 4.5\ c) \tag{13}$$

The effective fibre diameter to use is obtained from another relationship due to Davies [22]

$$D_e = \left[\frac{70 \mu L c^{1.5}\ (1+52c^{1.5})}{\Delta P} \right]^{1/2} \tag{14}$$

which requires that the pressure drop and filter thickness already be known. Note that the Kuwabara cell model (62), (63), (64), (65), Chapter 5, is not used.

A test of this model against some experimental data showed that while it underestimated the grade-efficiency somewhat, it followed the trend with d_{p_i} down to about 0.3 μm. Below that particle size the model failed to predict the increase in grade-efficiency due to the increasing importance of Brownian diffusion. Further attempts to check this model against experimental performance data revealed that it could not account for the effect of the humidity of the gas nor for the behavior of different types of fabrics.

B. System Analysis Model

Because of all of the complexities mentioned above, the design of fabric filters (baghouses) is best carried out by resort to empiricism and experience rather than a mathematical model for grade-efficiency. For this purpose the System Analysis Chart [23], Fig. 2 is useful to coordinate all of the information involved.

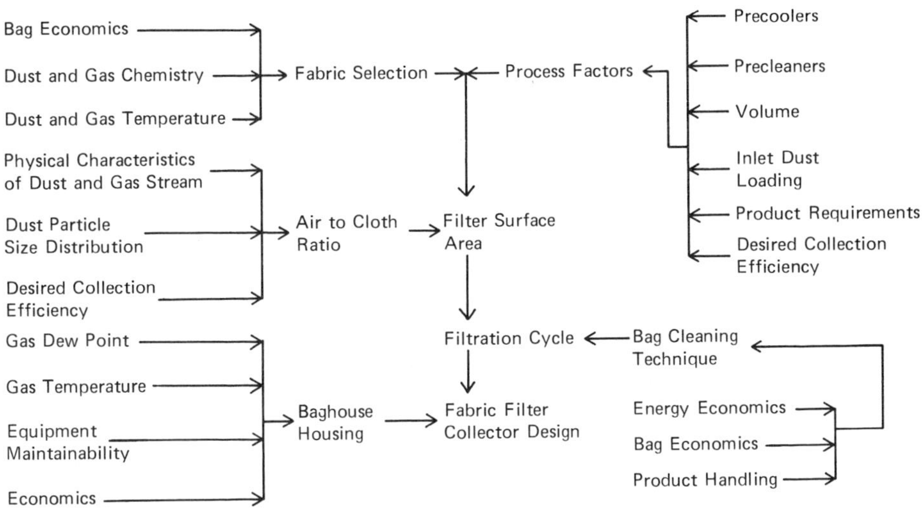

FIG. 2. System analysis for fabric filter design.

This chart shows that two major decisions must be made: selection of an appropriate fabric and determination of the "air-to-cloth ratio." These coupled with the process factors determine the total area of filter surface required, hence the number of bags and size of the baghouse. Subsequently the cycle of filtering and cleaning must be determined, along with the corresponding cycling of pressure drop.

1. Air-to-Cloth Ratio

This term is defined as the volumetric flow rate (Q) of the gas (e.g., acfm) divided by the total filter surface (A) of all bags (sq.ft). It is obviously the average lineal approach velocity $v_o = Q/A$. Since the value of Q is a given process factor, selection of v_o as a design parameter determines the total filtering surface required. It is desireable to have v_o as large as possible, consistent with attaining desired performance, in order to keep A as small as possible.

As shown in the design example for a fibre filter above, and as would be true in use of the Fraser-Foley or similar model, v_o affects η_{Ti} and ΔP. If capture theory applies, an increase in v_o should increase η_{T_i} for larger d_{p_i} because it will increase the effectiveness of the II mechanism. On the other hand, for very fine d_{p_i} when BD is the predominant mechanism the opposite effect would be predicted.

However, experience shows that one cannot develop a grade-efficiency relationship from classical capture theory. The mass emission rate of particles from a fabric tends to be independent of particle size, and always to increase with v_o [19]. Excessive air-to-cloth ratios must always be avoided.

One is forced to select v_o on the basis of case studies showing collection efficiency attained with a specified kind of dust using a specified kind of fabric. Extensive tables of such data are available [2], [24] from which a case must be selected to correspond as nearly as possible to that for which the design is to be undertaken. Overall efficiencies of greater than 99% are customarily attainable at values of v_o usually ranging between 1-6 ft/min.

a. Fault Processes. Recent research by Leith and First [19] has elucidated the collection behavior of fabric filters in terms of the concept of "fault processes" instead of capture mechanisms. Penetration is attributed primarily to the filters' inability to retain collected particles, rather than inability to capture them initially. It is postulated that essentially all particles are captured in attempting to negotiate the tortuous passage through the composite cake/fabric layer. Those which penetrate do so because of either "seepage" or "pinhole generation."

Seepage refers to the movement of arrested particles from their deposition sites, by a kind of flushing action of the gas. As pressure drop builds up the fabric may be stretched and particles loosened so that the drag of the gas can move them. Particles may be loosened during the cleaning process. This is a form of re-entrainment, occurring relatively slowly.

Pinholes are observed to form in the surface of a filter cake, at the pores between yarns in a woven fabric, or at places where synthetic felt was needle-punched during manufacture. Agglomerates of dust collected above such a hole may break loose and slip through producing coarser particulate in the emission than was in the feed! High localized gas velocity through a pinhole may prevent it from "healing" or becoming sealed over by additional dust. Instead anthill-shaped deposits surround the pinhole due to an inertial deposition as gas streamlines bend to pass through the hole.

The penetration of dust by seepage and pinhole formation may be lumped together as the total emission mass flux of dust which issues from the filter. This is found to increase with v_o, to be relatively independent of particle size in the feed, and to decrease markedly just after cleaning but subsequently to increase gradually and level off at a value corresponding to v_o. Emission flux continues from a deposited cake even though no new dust is being fed to the filter. It varies significantly throughout the filtering cycle for an individual bag, but there is an average steady value for a baghouse.

The goal of reducing the size of baghouses by operating at higher air/cloth ratios can only be approached by learning how to reduce the total emission flux. If this can be done fabric filters can approach being perfect collectors even of fine particles.

2. Pressure Drop

The flow through a fabric filter (cloth plus residual or deposited dust) follows Darcys' law, (6) above, so that $\Delta P \propto v_o$. The proportionality factor includes all the properties of the dust layer and fabric, as well as the effect of temperature (primarily upon μ). Three cases must be recognized: ΔP_c: a clean, unused fabric; ΔP_r: a used fabric containing a residual dust deposit, after cleaning; ΔP_f: fabric with increasing dust layer being deposited as filtering proceeds. For a clean fabric

$$\Delta P_c = K_c \mu v_o \tag{15}$$

K_c is determined by measuring the value of v_o for which $\Delta P = 1/2$ inch H_2O, for air flow at 20°C. This v_o is called the "permeability" of the fabric, and tables [2] of such values are available. While

ΔP_c is usually a negligible part of the total ΔP in operation, relative values of K_c, which depend only upon the composition and structure of the fabric, are useful in comparing fabrics.

For the residual after cleaning

$$\Delta P_r = K_r \mu v_o \tag{16}$$

when K_r is characteristic of the cloth/dust system and also of the method of cleaning and cycling. It is assumed that once a cloth is "broken in" by a few cycles of use an equilibrium of residual dust in the fabric will be reached. Thereafter K_r will have the same constant value at the beginning of each cycle. $\Delta P_r >> \Delta P_c$ and K_r automatically includes the effect of K_c.

For the filtering process, ΔP_f increases as the dust layer builds up in direct proportion to the weight of dust collected. A useful formulation is

$$\Delta P_f = K_f \left(\frac{c v_o t}{7000} \right) \mu v_o \tag{17}$$

when c = grain-loading of dust, t = time of filtration and 100% collection efficiency is assumed. The quantity in parenthesis is proportional to the thickness L as in (6) above.

The total pressure drop during filtration builds up linearly with the time:

$$\Delta P_{filt} = \Delta P_r + \Delta P_f = \mu v_o \left(K_r + \frac{K_f c v_o}{7000} t\right) \tag{18}$$

provided v_o is kept constant. Values of K_r and K_f, specific for each case, may be taken from recorded case studies. They are obtained from a plot of ΔP_{filt} vs t. The slope of the plot is proportional to K_f while extropolation to t = 0 gives an intercept from which K_r may be calculated. The values of the K's are found to depend upon humidity of the gas, as well as temperature of operation. Billings [18] gives a number of specific equations for various situations.

3. Design Procedure

a. Selection of Fabric. The first step in design is to select the fabric. This must be done with due consideration for a number of factors: ability of fabric to collect the particular dust, ability to withstand the corrosive and erosive effects of gas and dust, ability to withstand elevated temperature, suitable mechanical strength, cost, etc. These are all considered in the light of relevant case history.

The temperature and humidity of the gas are major considerations. Each type of fabric has a limitation on the maximum temperature at which it may be used. Since many filter applications are appropriate for hot gases, a pre-cooling system must usually be designed to precede the baghouse. As cooling proceeds the dew-point of the gas must be kept in mind. Condensation of moisture in the baghouse cannot be tolerated because of the "muddy" cake which would result. It is desirable to keep the gas temperature at least 50°F above the dew-point. Sometimes it is necessary to insulate the baghouse in order to prevent further natural cooling.

Some representative fabric characteristics are given in Table 1.

b. Pre-cooling System. Generally there are three methods of gas cooling to be considered when a pre-cooling system is required:

- cooling by dilution with ambient air;
- cooling in a heat-exchanger, or waste-heat boiler;
- cooling by evaporation of a water spray into the gas.

Each method has its advantages and disadvantages.

Cooling by dilution with ambient air is simple and relatively inexpensive. However, the greater the cooling required, the more ambient air is used and consequently there is a large net increase in volume of gas to be filtered. This increases the filtering surface area and the size and cost of the baghouse.

Cooling by means of some sort of heat exchanger will require the additional equipment for the purpose. This may be as simple as an extended surface of ductwork to provide for convective and radiative

Table 1

Filter Fabric Characteristics

Fibre	Operating exposure °F		Supports combustion	[a]Air permeability cfm/ft^2	Composition	[b]Resistance Rating to Abrasion	Mineral acids	Organic acids	Alkali	Cost rank
	Long	Short								
Cotton	180	225	yes	10-20	Cellulose	G	P	G	G	1
Wool	200	250	no	20-60	Protein	G	F	F	P	7
Nylon[d]	200	250	yes	15-30	Polyamide	E	P	F	G	2
Orlon[d]	240	275	yes	20-45	Polyacrylonitrile	G	G	G	F	3
Dacron[d]	275	325	yes	10-60	Polyester	E	G	G	G	4
Polypropylene	200	250	yes	7-30	Olefin	E	E	E	E	6
Nomex[d]	425	500	no	25-54	Polyamide	E	F	E	G	8
Fiberglass	550	600	no	10-70	Glass	P-F	E	E	P	5
Teflon[d]	450	500	no	15-65	Polyfluoroethylene	F	E	E	E	9

[a] cfm/ft^2 at 0.5 in. W.G.

[b] p = Poor, F = Fair, G = Good, E = Excellent

[a] Cost rank, 1 = lowest cost, 9 = highest cost.

[d] Dupont registered trademark.

cooling to the surrounding air. However, transfer rates will be low and the surface required large. If there is a great amount of heat to be removed it may be economically attractive to recover it, for example, in a waste heat boiler where there is a use for process steam.

Evaporative cooling will involve a rather simple spray system. The quantity of water required will be relatively small, hence its vapor will not increase appreciably the total volume of gas to be filtered. However, it will increase the humidity of the gas and the resulting increase in the dew-point will have to be taken into account.

Calculations for each of the methods of cooling will be illustrated in the design examples below.

c. Total Filter Surface. As soon as the fabric is selected, the appropriate air/cloth ratio may be determined. The actual volume flow rate (Q) of the gas to the filter may be calculated when the temperature of operation and method of cooling is fixed. Then A is calculated from $A = Q/v_o$.

A standard bag size (diameter by length) is also selected from manufacturers specifications. The number of bags then is equal to A divided by (πDL). The bags are usually divided up and grouped into compartments so that a compartment may be shut off-stream for cleaning or repairs without disturbing the filtering operation. For this reason it is usual to allow at least one compartment of excess filter surface above that calculated.

The temperature and humidity of gas at conditions expected to obtain in the baghouse must also be checked against the dew-point of the gas.

d. Determination of Filtering/Cleaning Cycle. The length of the filtering period is set on the basis of the changes in pressure drop and flow rate which may take place as the cake build-up proceeds. This must be done with regard to the characteristics of the fan being used to move the gas through the system, and to the total power consumption.

The fan must provide not only the ΔP of the filter as given by (18), but also a ΔP_{duct} for the ductwork of the system. The ΔP_{duct} is proportional to Q^2 as the gas is always in turbulent flow. The total ΔP_{sys} at any instant t is therefore of the form

$$\Delta P_{sys} = AQ + BQ^2 \tag{19}$$

and ΔP_{sys} must be provided by the fan.

It is characteristic of fans to have an operating relationship between ΔP_{sys} and Q which may be represented by

$$\Delta P_{sys} = C\ (1-Q^2/Q^2_{max})^{1/2} \tag{20}$$

in the operating range. These relationships (19) and (20) interact as shown in Fig. 3.

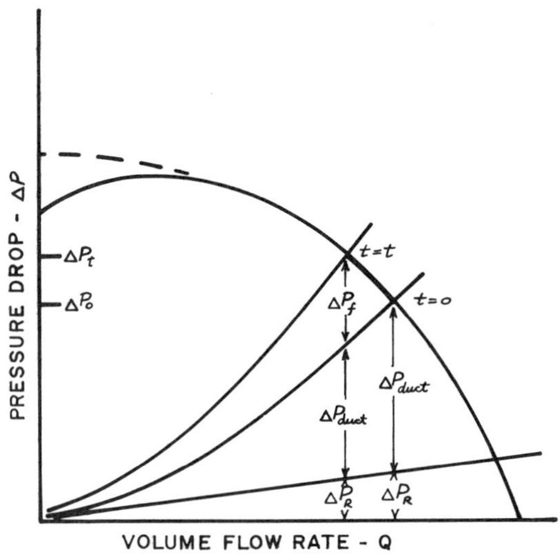

FIG. 3. Variation of pressure drop with flow rate and time in fabric filter cycle. Equations (19) and (20)

At the end of cleaning and beginning of filtering ($t = 0$) $\Delta P_{sys} = \Delta P_r + \Delta P_{duct}$ and the fan delivers flow rate Q_o. As filtration proceeds added resistance enters the system, increasing ΔP_{sys} by ΔP_f. The fan cannot continue to deliver Q_o, so the flow rate decreases along the fan curve. As it does so ΔP_{duct} and ΔP_r decrease somewhat, as ΔP_f continues to build up. The process moves along the fan curve from the point marked $t=0$ in the direction of the arrow toward any later time t. If all of the constants in (18), (19), and (20) are known this path may be calculated, and the amount of cake collected at t, as related to the ΔP_{sys}, may be determined.

The length of the filtering period (t) is usually determined on the basis of an allowable upper limit to ΔP_{sys}, based upon experience showing when an excessive value is reached. As a first approximation, if the fan curve (20) is not known Q_t may be taken as constant at Q_o.

The above remarks apply to a single bag, or to one compartment of bags operating together. For the complete analysis of performance of a number of compartments operating sequentially, see Crawford [25]. He also gives a detailed discussion of fan characteristics.

C. Design Example

An open-hearth steel plant operates six furnaces each having a capacity to produce 320 tons of finished steel in a nine-hour operating period. The charge to the furnace is 7% over the finished steel weight. The oxygen lance [26] is used, for a maximum of 30 min. each hour. Source emission tests indicate that the emission rate maximum (while lance is used) is 7.2 lbm/hr per ton of melt. At the same time the effluent from one furnace is equivalent to 62,000 std cfm (dry, at 70° F and 29.92" Hg) and leaves the duct at 1350° F. It is desired to limit the emission to 0.010 grains/acf at 450-600°F and 8% humidity by volume. The feasibility of using a fabric filter system is to be considered.

A cursory search for relevant case studies reveals that fabric filters have been used in such an application. The work of Herrick [26] is pertinent and may be drawn upon for this study. Other data may be found in [2] and [23].

Basic performance calculations are done on the basis of one furnace:

$$\text{Stack flow rate } Q = 62{,}000 \ \frac{1350 + 460}{70 + 460} \times \frac{1}{(1-0.08)} = 230{,}150 \text{ acfm}$$

$$\text{Emission flux rate } E = 7.2 \times 320 = 2304 \text{ lbm/hr maximum}$$

$$\text{Maximum grain loading} = \frac{2304 \times 7000}{230{,}150 \times 60} = 1.17 \text{ gr/acf}$$

Maximum efficiency required ~ minumum penetration allowed

$$P = 0.010 \times \frac{450 + 460}{1350 + 460}/1.17 = 4.29 \times 10^{-3}$$

$$\eta_M = 0.9957$$

Although conditions fluctuate during this process, the design should be based upon the "most difficult" set of conditions and not upon average conditions.

Selection of Fabric: Because of the high temperatures involved a pre-cooling system will be required, but a fabric to operate as hot as possible should be sought. It must also be established that the fabric selected can collect this kind of dust at 99.6% efficiency.

No data on the particle size and composition of this dust is given, but a good deal may be inferred from the literature. The material is primarily iron oxides, with small amounts of impurities, 45-50% in the 0-5 μm size range.

The Filter Handbook [2], Table 2.31, lists Fe_2O_3 cupola dust, 0.7 gr/acf, collected by fibre glass filter using shake cleaning, at

450°F, with an air/cloth ratio of 2.1 ft/min. The Particulate Control Document [23], Sec. 4.7 and 5.4, indicates that fibre-glass bags may be used at 500°F, using reverse flux cleaning and sonic horns, with a ratio of 2 ft/min on a grain loading as high as 20 gr/acf. Table 1 shows fiberglass operating steadily at 550°F, and intermittently up to 600°F. The Herrick work [26] confirms these statements, indeed is probably the source of them.

We may therefore confidently and conservatively make the following selection: Fabric: Fibre-glass bags; Operating temperature: 500°F; Air/Cloth Ratio: 2 ft/min; Efficiency and grain loading: as required.

Gas Cooling: A system to cool 230,150 acfm of gas at 1350°F to 500°F must be designed. For the sake of illustration let us look at all of the possible cooling systems discussed above. The gas is essentially N_2 and O_2 with some CO_2, CO, and 8% vol. H_2O. It may be taken to have the properties of air in lieu of precise data on its composition: ρ_f = 0.0808 lbm/ft^3, C_p = 0.25 Btu/lbm°F. The cooling required is as follows:

Mass of gas = 230,150 x 0.0808 x 492/1810 = 5050 lbm/min

Mass of dust = 2304/60 = 38.4 lbm/min

Heat removal rate = 5090 x 0.25 x (1350-500) = 1.082 x 10^6 Btu/min

Dilution with ambient air at say 100°F will require: 1.082 x 10^6/0.25 (500-100) = 10,816 lbm/min which will add 10,816 x 960/0.0808 x 492 = 261,200 cfm of volume flow to the gas at 500°F which would enter the baghouse. The total Q for the baghouse would thus be:

Q = 261,200 + 230,150 x 960/1810 = 383,300 acfm

against a volume of only 122,100 acfm unduluted. Dilution will more than triple the filtering surface required, which is clearly excessive.

A convective/radiative heat exchanger conservatively estimated to have an overall heat transfer coefficient of 10 Btu/hr ft^2°F, operating with a log-mean temperature difference of (1250-400)/(ln 1250/400) = 746°F will require a

$$\text{Heat Exchange Surface} = \frac{1.082 \times 10^6 \times 60}{746 \times 10} = 8700 \text{ ft}^2$$

This corresponds to a very large heat exchange unit. However, the volume of gas to be filtered would remain at 122,1000 acfm.

For an evaporative spray cooler it may be assumed that the water will remove 1000 Btu/lbm evaporated. This will require 1.082 1.082 x 10^6/1000 = 1082 lbm/min or 130 gal/min of water. This will add 1082 x 38.8 = 42,000 ft^3/min of vapor volume to the gas at 500°F, making Q = 42,000 + 122,100 = 164,100 acfm. While this would increase the filter surface by about 35%, an equally significant consequence would be the effect on the humidity. This would increase from 8% by volume to (42,000 + 0.08 x 122,100)/164,100 or 31.5%. The absolute humidity will be approximately

$$\text{A.H.} = \frac{(122{,}100 \times 0.08 + 1082)/38.8}{5050 - (122{,}100 \times 0.08)/38.8} = 0.278 \text{ lbm } H_2O/\text{lbm dry gas.}$$

Examination of a psychrometric chart shows that this will correspond to a dew point of about 150°F.

A final selection among these cooling processes cannot be made without a detailed economic analysis, which should consider the possibility of a waste-heat boiler as well. However, on the basis of the above calculations only, it seems that the evaporative cooler may be the most practical and the baghouse will be sized accordingly.

Baghouse Design: Taking v_o = 2 ft/min, Q = 164,000 acfm, the total filter surface required will be A = 82,050 ft^2. A baghouse configuration may be based upon bags each 11½" diameter by 34 ft. long, 80 bags per compartment [26]. This will correspond to 82,050 x 0.958 x 34 x 80/π = 10.02 compartments. A reasonable design might use 11 compartments, assuming one at a time is off-stream for cleaning. Gas will enter the baghouse at 500°F and should not leave below 200°F.

Filtering Cycle: Some relevant data on the pressure drop constants for this kind of a system may be found in sources mentioned above.

From the Handbook, Table 2.31 [2], for iron oxide cupola dust on fibreglass the following values were obtained: $\Delta P_r/v_o$ = 2.9 in. H_2O per ft/min; $\Delta P_{filt}/v_o$ = 4.3 in H_2O per ft/min, maximum; (K_{fu}) = 121 in. H_2O x ft x min/lbm all at v_o = 2.1 ft/min, c = 0.7 gr/ft^3, and 450°F.

Equation (18) may therefore be set up to read

$$\Delta P_{filt} = 2.9\ v_o + 121\ c\ v_o^2\ t/7000$$

and for the present case (v_o = 2 ft/min at 500°F) in which the value of c will depend upon the actual gas flow according to the method of pre-cooling used. The values of c, and corresponding equations for ΔP_{filt} are as follows.

Dilution: $$c = 1.17 \times \frac{230,150}{383,300} = 0.703 \text{ gr/acf}$$

$$\Delta P_{filt} = 5.8 + 4.86 \times 10^{-2}\ t$$

Exchanger: $$c = 1.17 \times \frac{1810}{960} = 2.21 \text{ gr/acf}$$

$$\Delta P_{filt} = 5.8 + 0.153\ t$$

Evaporative: $$c = 1.17 \times \frac{230,150}{164,100} = 1.64 \text{ gr/acf}$$

$$\Delta P_{filt} = 5.8 + 0.113\ t$$

If we accept the value of $\Delta P_{filt}/v_o$ = 4.3 in H_2O as the maximum tolerable, then ΔP_{filt} = 8.6 in H_2O and the length of filtration time is, in each case

Dilution: t = 2.80/0.0486 = 57.6 min

Exchanger: t = 2.80/0.153 = 18.3 min

Evaporative: t = 2.80/0.113 = 24.8 min

In lieu of information about a specific fan performance curve (Fig. 3) these calculations assume that Q remains constant during the filtering cycle. The ΔP_{duct} has not been considered either.

All of the above constitutes only a preliminary survey of the problem. But it is sufficient to indicate that a baghouse filter may be feasible for the process and warrants further study and consideration.

REFERENCES

1. C.N. Davies, *Air Filtration*, Academic Press, London, 1973.

2. C.E. Billings and J. Wilder, *Handbook of Fabric Filter Technology*, Vol. I PB 200-648, Vol. II PB 200-649, U.S. Public Health Service, 1970.

3. H.C. Yeh and B.Y.H. Liu, *Aerosol Sci.*, 5: 191 (1974).

4. H.C. Yeh, Ph.D. Thesis, U. of Minnisota, (1972).

5. H.C. Yeh and B.Y.H. Liu, *Aerosol Sci.*, 5: 205 (1974).

6. L.A. Clarenburg and J.F. Vander Wal, *Ind. Eng. Chem. Proc. Des. Devel.*, 5: 110 (1966).

7. D.C. Freshwater and J.I.T. Stenhouse, *A.I.Ch.E. Journ.*, 18: 786 (1972).

8. M.W. First and W.C. Hinds, *Journ. Air Poll. Control Assoc.*, 26: 119 (1976).

9. C.N. Davies, *Proc. Inst. Mech. Eng.*, 1B: 185 (1952).

10. L.A. Clarenburg and R.M. Werner, *Ind. Eng. Chem. Proc. Des. Dev.*, 4: 293 (1965).

11. T.E. Wright, R.J. Stasny and C.E. Lapple, *WADC Tech, Rep. 55-457*, Donaldson Co., Inc. (1957) PB-131, 570.

12. S.C. Yung, R.G. Patterson, S. Calvert, and D.C. Drehmel, Paper 77-32.6, Air Pollution Cont. Assn., 70th Annual Meeting, Toronto, 1977.

13. S. Jackson and S. Calvert, *A.I.Ch.E. Journ.*, 12: 1975 (1966).

14. S. Ergun, *Chem. Eng. Prog.*, 48: 89 (1952).

15. C.H. Black and R.W. Boubel, *Ind. Eng. Chem. Process Des. Dev.*, 8: 573 (1969).

16. D. McCarthy, A.J. Yankel, R.G. Patterson, M.L. Jackson, *Ind. Eng. Chem. Process Des. Dev.*, 15: 266 (1976).

17. A.R. Figueroa and W. Licht, *A.I.Ch.E. Sym. Series* (1977) in Press.

18. C.E. Billings, *A.I.Ch.E. Symp. Ser. No. 137*, 70: 341 (1974).

19. D. Leigh and M.W. First, *Journ. Air Poll. Control Assoc.*, 27: 534, 636, 754 (1977).

20. M. Fraser and G.J. Foley, Paper 74-99, Air Poll. Control Assoc., 67th Annual Meeting, Denver, 1974.

21. C.Y. Chen, *Chem. Rev.*, 55: 595 (1955).

22. C.N. Davies, *Proc. Instn. Mech. Engrs. London*, 1B, 185 (1952).

23. *Control Techniques for Particulate Air Pollutants: AP-51*, U.S. Dept. of Health, Education and Welfare, 1969.

24. W. Strauss, *Industrial Gas Cleaning, 2nd Ed.*, Pergamon Press, New York, 1975.

25. M. Crawford, *Air Pollution Control Theory*, Sec. 5-8, 10-8, McGraw-Hill, New York, 1976.

26. R.A. Herrick, J.W. Olsen, F.A. Ray, *Journ. Air Poll. Control Assoc.*, 16: 7 (1966).

27. N. Kimura and K. Iinoya, *Kagaku Kogaku*, 23: 792 (1959).

CHAPTER 9

WET SCRUBBING

I. SCRUBBER CHARACTERISTICS

A. General Features

Scrubbing a particle-laden gas by bringing it into contact with a liquid can be an effective means of removing small particles. It also provides the possibility of dissolving gaseous components from the stream, thus providing for two kinds of purification simultaneously. The particulate matter in a wet state and the spent scrubbing liquor must be disposed of, and this obviously presents special problems not encountered in dry methods of collection.

There are many kinds of scrubbing devices. In an extensive study Calvert [1] has classified them into ten types. A handy summary of their designs and performance is also available [2]. However for purposes of modelling, these may be grouped broadly into two main classes: (a) those in which an array of liquid drops (sprays) form the collecting targets; (b) those in which wetted surfaces of various kinds are the collecting medium. In the first class are to be found preformed sprays, gas atomized sprays, and self-induced sprays. In the second class are a variety of plate towers and packed towers, including massive packing, fibrous packing, moving bed packing and irregated fibre filters. The treatment in this chapter will be devoted to the first class.

One unique feature of wet collectors is the necessity for providing energy to move the liquid through the device as well as the gas. The total scrubber power required is an important parameter. It is well established that the higher the scrubber power, per unit volume of gas treated, the lower will be the cut diameter of the particles emitted. Calvert [2] uses this to correlate and compare the performance of the several kinds of devices. He shows that cut diameters of about 1 micron may be obtained with gas atomized spray devices, at about 1 hp per 1000 acfm of scrubber power. To obtain a cut diameter of 0.5 micron will require over 6 hp/1000 acfm. As shown by the sample calculation of Example 5 in Chapter 3, this is substantially higher than typical values for other devices. For the collection of sub-micron particles, the scrubber is a "high energy" device.

B. Spray Scrubbers

The term spray scrubber will be used to refer to a device in which the liquid is broken into drops, roughly in the range 100-1000 μm diameter, which are introduced into the particulate-laden gas stream. There the array of moving drops becomes the set of targets (regarded as spherical) upon which the particles may be collected by any of the mechanisms of aerodynamic capture. Of special interest are the possibilities of capture by diffusiophoresis and thermophoresis, as well as by "electrified" drops. (See Chapter 5)

Models for collection are based upon the concept of single-target efficiency for an individual sphere, by assuming that this efficiency may be calculated from the parameters of aerodynamic capture, as presented in Chapter 5. Inertial impaction (II) remains an important mechanism of capture for super-micron particles. For this to occur there must be relative motion between drops and particles. This would also be true of direct interception (DI). For any of the diffusion mechanisms or electrostatic forces, however, relative motion is not required except to bring the particles sufficiently close to the drops so that these kinds of forces may become effective. They are, of course, of primary importance for the sub-micron particles.

Relative motion is provided in all types of scrubbers. The principle kinds are:

(a) Pre-formed spray: drops are formed by atomizer nozzles and sprayed into gas stream: (i) Counter-current gravity tower: drops settle vertically against rising gas-particle stream; (ii) Cross-current tower: drops settle through horizontal gas stream; (iii) Co-current: spray is horizontal into horizontal gas stream; (iv) Ejector venturi: liquid sprayed at high pressure into converging section ahead of throat pulls gas in as well.

(b) Gas-atomized spray: liquid is introduced more or less in bulk ahead of "venturi" throat and shattered into drops by high-velocity gas: (i) Spray venturi: liquid squirted in around periphery of venturi, or flows over weir into gas; (ii) Annular orifice: liquid and gas flow in parallel into annular space between a movable disk and venturi walls, ahead of throat; (iii) Rod bank: liquid and gas flow in parallel into constricted space formed between rods arranged parallel to each other in a row. From the point of view of basic principles it is better to treat all of the venturi types together as a class, whether pre-formed (a, iv) or gas atomized (b,i; b,ii; b,iii).

II. NON-VENTURI SCRUBBERS

A. Gravity Chambers

1. Counter-Current Flow

Liquid is sprayed in at the top of a tower through atomizing nozzles. Drops then fall freely at their terminal settling velocities relative to the gas. The gas/particle stream is introduced at the bottom, flows upward against the rain of drops and clean gas issues from the top. Liquid containing particulate collects in a pool at the bottom and is pumped out for treatment to remove the solids. Cleaned liquid is usually recirculated to the sprays. The tower is simply an empty chamber.

The approach to modelling the collection process is similar to that used for fibrous filters. Consider a cross-section of such a tower of differential thickness as shown in the sketch:

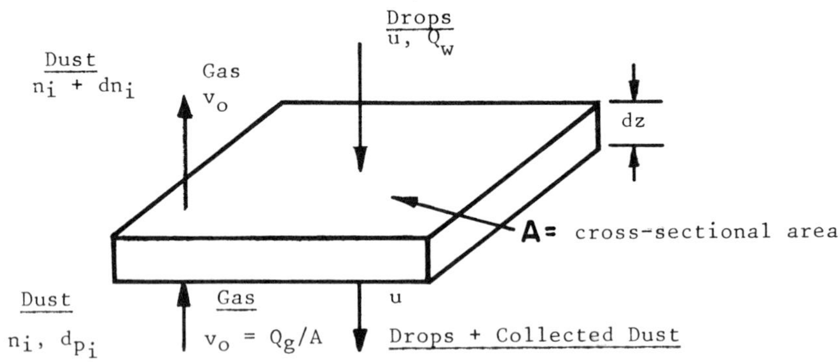

The model is developed on the assumptions that: (1) The drops are all of the same size, diameter = D, and are falling at their terminal velocity u_{s_D} relative to the gas. Thus, the velocity of the drops relative to the tower is $u = (u_{s_D} - v_o)$. (2) The drop concentration is sufficiently dilute that each drop may be regarded as acting independently as though it were present alone. (3) The total quantity of scrubbing liquid, flowing at volume rate Q_ℓ is distributed uniformly over the cross-sectional area A. The total number of drops entering the section per unit time is equal to $6Q_\ell/\pi D^3 A$. (4) The individual dust particles are being carried upward in the tower at a velocity of $(v_o - u_{s_i})$ relative to the tower, where u_{s_i} = terminal settling velocity of dust particles of ith grade (in many cases $u_{s_i} << v_o$, and $u_{s_i} << u_{s_D}$ also). (5) The concentration of dust of grade i is uniform across the section and equal to n_i (number of particles/volume of gas) i.e., there is complete lateral mixing of uncollected particles across any level. (6) According to the definition of single target total efficiency η_{T_i}, each drop collects, per unit time, a number of particles equal to: $\eta_{T_i}\,(u_{s_D} - u_{s_i})n_i\,\pi D^2/4$, wherein $\eta_{T_i} = f\,(Stk_i, Re_D, Pe_i, R_i$, etc.). (7) The time required for a drop to fall through the section of thickness dz is equal to $dz/u = dz/(u_{s_D} - v_o)$.

The total collection of particles taking place in the section is

$$-dn_i \frac{Q_\ell}{A} = \frac{6Q_\ell}{\pi D^3 A}\,\eta_{T_i}\,\frac{\pi D^2}{4}\,(u_{s_D} - u_{s_i})\,n_i\,\frac{dz}{u} \qquad (1)$$

or

$$\eta_{s_i} = 1 - \exp\left[-\frac{3}{2}\,\frac{(u_{s_D} - u_{s_i})}{(u_{s_D} - v_o)}\,\frac{Q_\ell}{Q_g}\,\frac{H}{D}\,\eta_{T_i}\right] \qquad (2)$$

Integrating on the assumption that the quantities on the right are constant and independent of z, given the overall efficiency:

$$\int_{n_{o_i}}^{n_i} \frac{dn_i}{n_i} = \frac{3}{2}\int_0^H \frac{(u_{s_D} - u_{s_i})}{(u_{s_D} - v_o)}\,\frac{Q_\ell}{DQ_g}\,\eta_{T_i}\,dz$$

where H = total height of tower. This is a grade-efficiency relation because η_{T_i} depends upon d_{p_i}.

Alternate forms of (2) which are useful may be obtained in dimensional forms by: (a) letting $L \equiv Q_\ell$, expressed in gallons liquid per unit time, and $G \equiv Q_g$, expressed in thousands of ft^3 of gas per unit time, then expressing D in microns, H in feet:

$$\eta_{s_i} = 1 - \exp\left[-71\,\frac{(u_{s_D} - u_{s_i})}{(u_{s_D} - v_o)}\,\frac{L}{G}\,\frac{H}{D}\,\eta_{T_i}\right] \qquad (3)$$

or (b) letting Ω represent the total surface of drops per unit volume of gas, expressed in square feet/cubic foot, and the total height H in feet, $\Omega = 6Q_\ell/DQ_g$ and

$$\eta_{s_i} = 1 - \exp\left[-\frac{1}{4} \times \frac{(u_{s_D} - u_{s_i})}{(u_{s_D} - v_o)} \Omega H \eta_{T_i}\right] \tag{4}$$

In case the drops are not all of the same size it is customary to use the Sauter mean drop diameter to calculate Ω in (4), or simply to take D as the Sauter mean in (3). It is important to know the drop size distribution produced by the spray nozzles in this case. A more rigorous approach would be to divide the drop sizes into intervals, represented by D_j, and to determine a value of $\eta_{T_{i,j}}$ for each of the combinations of (d_{p_i}, D_j) pairs. Then equation (2) would have to be applied to each pair, and the total collection for all pairs summed.

The role of the drop size may be explored in part through its effect upon η_{T_i}. In the usual range of operation for gravity spray towers it is found II is the predominant mechanism of capture. Using a relationship recommended by Calvert [3]

$$\eta_{T_i} = [Stk_i/(Stk_i + 0.35)]^2$$

and

$$Stk_i = \frac{\rho_p \, d_{p_i}^{\,2} \, (u_{s_D} - u_{s_i})}{18\mu \qquad D} = \tau_i \frac{u_{s_D}}{D} \tag{5}$$

so that

$$\eta_{T_i} = \left[\frac{\tau_i \, u_{s_D}}{\tau_i \, u_{s_D} + 0.35}\right]^2 \tag{6}$$

For small drops $u_{s_D} \propto D^2$ (Stokes law) and η_{T_i} increases as D increases. At intermediate sizes $u_{s_D} \propto D$ and η_{T_i} is constant. For larger sizes u_{s_D} increases less rapidly than D and η_{T_i} decreases. Hence there is a value of D for which η_{T_i} is a maximum. Stairmand [4] has calculated that the peak value of η_{T_i} occurs at about $D \approx 600\mu m$ regardless of particle size, is larger for larger particles, and is a rather flat peak extending for two or three hundred microns on either side of $D \approx 600\mu m$.

The total effect of drop size on overall efficiency involves the interplay of u_{s_D}, η_{T_i}, and D as they appear in (2) or (3). Since η_{T_i} is relatively constant with D over a central range in which $u_{s_D} \propto D$, the net effect on the factor $u_{s_D}\, \eta_T\, (\mu_{s_D} - v_o)\, D$ is to make it a maximum at the low D end of this range, say in the neighborhood of 300-400 μm.

2. Cross-Flow

Proceeding in an analogous fashion, a model may be developed for the case of horizontal cross-flow of particle-laden air through a vertical settling spray. Similar assumptions are made, except that $v_o = 0$ in the vertical direction. The result is the same as (2) with $v_o = 0$. Then, also taking $\mu_{s_i} >> \mu_{s_d}$:

$$\eta_{s_i} = 1 - \exp\,(-\,3\,Q_\ell\,H\,\eta_{T_i}/2\,Q_g\,D) \tag{7}$$

The same equation as above may be used for calculating η_{T_i} but the value of Stk_i must be based upon v_o: $Stk_i = \tau_i\, v_o/D$. Now the ratio η_{T_i}/D in (7) clearly has no maximum value with respect to D, but continues to increase as D decreases. Alternates to (7) in forms analogous to (3) or (4) may easily be written.

The case of a spray nozzle operating horizontally in a duct with a co-current flow of gas, type a(iii), has been analyzed by Cheng [5]. He obtains equation (7), modified by the insertion of a factor λ ahead of Q_ℓ and replacement of H by a length ℓ. These modifications take into account the fact that the spray issues from the nozzle (assumed to be placed at the center of the duct of radius R) in a conical shape, with half-angle α, which does not fill the duct completely for some distance downstream, and that the spray drops have all settled to the bottom of the duct by a distance S ("projective distance" of the spray) downstream. Letting $r = S/R$, λ and ℓ are obtained:

$$\lambda = \frac{3r - 2 \cot \alpha}{2r^3 (1 - \cos \alpha)} \qquad (8)$$

$$\frac{\ell}{R} = r - \frac{2}{3} \cot \alpha$$

and S is estimated according to the exit velocity of drops from the nozzle v_D and their terminal settling velocity:

$$S \approx (v_D - v_o) R/u_{s_D} \qquad (10)$$

The model is shown to give good agreement with experiment, when η_{T_i} is based upon Stk_i in which the relative velocity is taken as the arithmetic mean of v_D and the final axial drop velocity, when the drop reaches the duct.

B. The L/G Ratio

These models introduce the liquid/gas flow ratio, which is of basic importance in the performance of all kinds of scrubbers. This ratio should be expressed fundamentally in terms of mass flow rate of liquid (M_ℓ) divided by mass flow rate of gas (M_g). For convenience however this is usually expressed in terms of the equivalent volumes taken at the actual conditions of entrance. Thus

$$\frac{M_\ell}{M_g} = \frac{\rho_\ell Q_\ell}{\rho_g Q_g}$$

For the air-water system at 20°C, $\rho_\ell/\rho_g = 1/1.205 \times 10^{-3} \approx 830$. It is also common practice to express Q_ℓ/Q_g in units of gallons of liquid per thousand actual cubic feet of gas, at entrance conditions. This is commonly called the L/G ratio. Experience shows that L/G will typically fall in a range of roughly 2-20 gal. per 1000 cu ft, or 0.27-2.7L/m^3. (1 gal/Mft^3 = 0.1337 L/m^3.)

This ratio is directly proportional to the number of drops per unit volume of gas: $6Q_\ell/\pi D^3 Q_g = 2.553 \times 10^8$ L/G D^3 drops/cm^3, where D must be in microns. The following tabulation shows the number of drops per cubic centimeter to be obtained at various values of L/G and D.

L/G gal/Mft^3	2	10	20
D = 100 μm	511	2553	5106
200 μm	64	319	638
400 μm	8	40	80
800 μm	2	5	10

It is clear that at combinations of low D and high L/G, the drops would be so crowded that collision and coalescence would be likely to occur. Assumption (2) of the model would not be valid. An operation attempted at too small D would automatically move toward a larger D. The feasible range of operating conditions is thus roughly delineated.

Example 1. Develop a set of grade-efficiency curves for a countercurrent gravity scrubber which is to remove particles of 2.60 gm/cm^3 density ranging in size from 0.6 μm to 20 μm. Consider L/G ratios of 5 and 10 gal/Mft^3 respectively, water drop sizes of 200, and 500 microns, and tower heights of 2 ft. and 10 ft. Assume the gas to be saturated with, and at the same temperature as the liquid. Discuss what differences there would be if the liquid is colder than the saturated gas. (See Chapter 5.) Take v_o at 0.20 u_{s_D}, and also at 0.50 u_{s_D}.

Figure 1 displays these grade efficiency curves for five cases. Case I may be regarded as the base case for which L/G = 10 gal/M ft^3, D = 500 microns, H = 2 ft, and v_o = 0.2 U_{s_D}. The other cases show that the grade efficiency is increased by: smaller drop size D (Case II: D = 200 μm); larger ratio v_o/u_{s_D} (Case III: v_o = 0.5 u_{s_D});

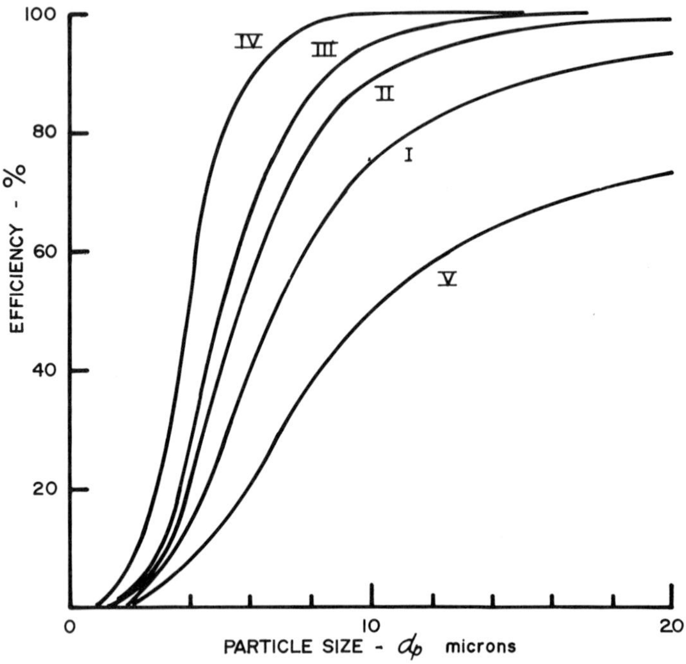

FIG. 1. Grade efficiencies for spray tower.

Case	L/G	D	H	v_o/u_{s_d}
I	10	500	2	0.20
II	10	200	2	0.20
III	10	500	2	0.50
IV	10	500	10	0.20
V	5	500	2	0.20

larger tower height H (Case IV: H = 10 ft.); smaller L/G (Case V: L/G = 5 gal/M ft^3). These results are typical: the greatest efficiency would be obtained (out of these possible values) by the combination D = 200 μm, $v_o = 0.5\ u_{sD}$. L/G = 10 gal/M ft^3, H = 10 ft. The calculations are straightforward using equation (3), (5), and (6) together with the methods given in Chapter 4 for determining u_{sD}.

C. Tower Design

The above example focuses attention on some of the questions arising in tower design. As in the case of other collectors discussed in previous chapters, the value of Q_g is a specified requirement.

However v_o, Q_ℓ/Q_g, and D may be selected by the designer. Selection of v_o determines the tower cross-sectional area $A = Q_g/v_o$. Obviously $v_o < u_{s_D}$, but the larger v_o the longer the residence time of drops in the tower and the greater the overall efficiency.

The height H of the tower may be determined by applying (3) to the particle size distribution in the same manner as was done for precipitators, and fibre filters according to equation (18), Chapter 3. An equation is set up to express the desired overall efficiency (see Chapter 7, page 293, and Chapter 8, page 310). This equation is then solved, by iteration, to give H. The terms in the equation depend upon the values selected for v_o, L/G, D, as well as the particle-size grade interval for g_i values.

Thus a range of possible tower sizes (A and H) exists in which all will be technically satisfactory solutions. The one combination to be selected must, as usual, be determined by an optimizing procedure. The applicability of the model is limited, as always, by the accuracy with which the conditions in a real scrubber are described by the set of assumptions made in deriving the equations.

The interplay among v_o, L/G, D, and scrubber geometry will be found to have a precisely analogous counterpart in the design of Venturi scrubbers. With obvious modifications the system chart Fig. 3 may be applied to any scrubber.

III. VENTURI SCRUBBERS

A. Basic Aspects

The essential feature of a Venturi scrubber is the presence of a constricted cross-section or "throat" through which the gas is forced to flow at high velocity. Drops introduced ahead of the throat at very low axial velocity are accelerated by the drag of the high speed gas. This provides the relative velocity needed for inertial and other collection mechanisms.

The cross-section may be either circular or rectangular. The circular shape is not used in larger sizes (say above 1 ft. diameter) because it is too difficult to get a uniform distribution of drops across the entire section. The rectangular shape is therefore used with a rather large aspect ratio, in order to provide for a short side across which a good distribution can be obtained. This especially is the case in type b(iii) the parallel rod bank.

The cross-sectional area of the throat A_t determines the maximum velocity of the gas: $v_t = Q_g/A_t$. Ahead of the throat there is a converging section in which acceleration of the gas occurs. Following the throat is a more gradually diverging section at the end of which the gas has returned to its initial velocity. The collection process may be essentially completed by the end of the throat. The throat length then would have to be sufficient to accomplish the desired collection. For the parallel rod bank, the drops are introduced upstream and the "throat" itself is very short. Some acceleration of the drops will occur prior to the rods and the final acceleration, relatively brief, in between the rods. Also, in this case there are a number of "venturis" operating in parallel.

The modelling of the Venturi must include the following features: (i) geometry of sections: throat velocity; (ii) introduction and acceleration of drops: drag on accelerating drops; (iii) relative velocity vs. throat length; (iv) drop size: preformed sprays, gas-atomized sprays; (v) the capture process: collection efficiency; (vi) the pressure drop: power required.

B. Modelling

1. Venturi Geometry

Figure 2 shows a typical outline for types a(iv), b(i), b(ii), with the various sections designated. The cross-sectional area, with reference to that of the throat is given by:

$$A = A_t\left(1 + \frac{z_2 - z}{r_t} \tan \beta\right) \text{ converging:} \quad z < z_2$$

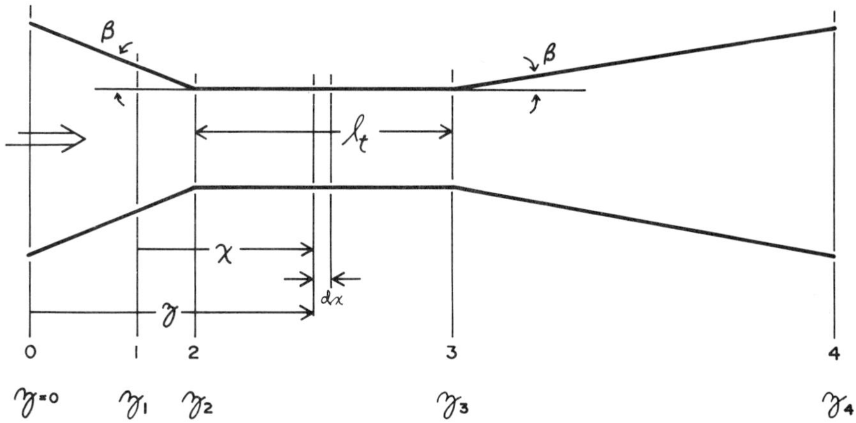

FIG. 2. Venturi geometry. 0-2 Converging Section, 2-3 Throat, 3-4 Diverging Section, 1 drops introduced; may coincide with 2; $s = z - z_1$.

$$A = A_t\left(1 + \frac{z - z_3}{r_t} \tan \beta\right) \text{ diverging: } z > z_3$$

$$A_t = \pi r_t^2 \text{ for circular sections}$$

$$A_t = 2r_t \cdot W \text{ for rectangular } (2r_t \text{ by } W) \text{ section}$$

Drops are introduced at z_1, which often nearly coincides with z_2. In case $z_1 < z_2$, acceleration of drops must be considered in the converging section as well as the throat.

The gas velocity v_g at any point is given by

$$v_g = M_g/\rho_g A = \frac{v_t}{1 + \dfrac{(z - z_2)}{r_t} \tan \beta} \tag{11a}$$

or

$$v_g = M_g/\rho_g A = \frac{v_t}{1 + \dfrac{(z - z_3)}{r_t} \tan\beta} \tag{11b}$$

where M_g is the mass flow rate of gas, taken as steady, and v_t is the gas velocity in the throat.

It is generally assumed that the flow through the venturi is incompressible, i.e., that ρ_g remains constant, independent of z. However, in some operations ΔP along the throat may be as high as 100 in. H_2O (≈ 3.6 lb/in^2). If the initial upstream pressure is relatively low, then it may be necessary to make a correction for compressible flow of the gas according to the standard venturi equations.

2. Introduction and Acceleration of Drops

Drops may be introduced into the gas stream in a variety of ways. The location of introduction will depend in part upon the method. The method will also be a predominant factor in determining the size and size distribution of the drops. These matters will be taken up below.

Consider an individual drop, introduced with some initial axial velocity, which may often be taken as zero. It will be accelerated according to Newton's Law, by the drag force exerted by the gas (See Chapter 5).

$$F_D = \frac{\pi D^3}{6} \rho_D \frac{dv_D}{dt} = C_{DA} \frac{\pi D^2}{4} \rho_g \frac{(v_g - v_d)^2}{2} \tag{12}$$

or

$$\frac{dv_D}{dt} = \frac{3}{4} C_{DA} \frac{\rho_g}{\rho_D} \frac{(v_g - v_D)^2}{D} \tag{13}$$

Replacing

$$\frac{dv_D}{dt} = \frac{dv_D}{dx} \frac{dx}{dt} = v_D \frac{dv_D}{dx}$$

the differential equation relating v_D to x (distance downstream from z_1) becomes

$$\frac{dv_D}{dx} = \frac{3}{4} C_{DA} \frac{\rho_g}{\rho_D} \frac{(v_g - v_D)^2}{v_D D} \tag{14}$$

Solution of (14) depends upon selection of an appropriate representation for C_{DA}. This question was raised in Chapter 4 when it was shown that in general $C_{DA} \neq C_D$. In the range of Reynold's Number encountered in venturis the results of Ingebo [6] may be applicable:

$$C_{DA} = 27 \, Re_D^{-0.84} \text{ where } Re_D = D\,(v_g - v_D)\,\rho_g/\mu$$

This relationship would be awkward to use, however, for when substituted in (14) it will introduce fractional powers which will make integration difficult. It has also been criticized on experimental grounds by Boll [7].

Another approach has been taken by Hollands and Goel [8], to represent C_{DA} in terms of C_{D1}, based upon conditions at z_1. Thus

$$C_{DA} = C_{D1} \left[\frac{Re_{D1}}{Re_D}\right]^{0.5} = C_{D1}\left[\frac{(v_{g1} - v_{D1})}{(v_g - v_D)}\right]^{0.5} \tag{14a}$$

Here C_{D1} is taken from the standard relationship for steady motion, as represented in Chapter 4, based upon $Re_{D1} = D\,(v_{g1} - v_{D1})\,\rho_g/\mu$. Substitution in (14), gives

$$\frac{dv_D}{dx} = \frac{3}{4} \frac{\rho_g}{\rho_D} \frac{(v_g - v_D)^{1.5}}{v_D D} C_{D_1} (v_{g_1} - v_{D_1})^{0.5} \tag{15}$$

Integration of (15) depends upon the conditions at drop introduction. For the case where the drops are introduced at the beginning of the throat by overflow through a weir or through nozzles, type b(i), $z_1 = z_2$, $v_{D_1} = 0$ at $x = 0$, and $x = \ell_t$ at the end of the throat. Then $v_g = v_1 = Q_g/A_t = v_t$, and (15) leads to

$$\frac{1}{v_t^{0.5}} \int_o^{v_D} \frac{v_D \quad dv_d}{(v_t - v_D)^{1.5}} = \frac{3}{4} \frac{\rho_g \ C_{D1}}{\rho_D \ D} \int_o^x dx \tag{16}$$

This has been integrated by Yung, Barbarika, and Calvert [9] to produce, in dimensionless form

$$u = {}^{v}D/{}^{v}t = 2\ [1 - X^2 + X\sqrt{X^2 - 1}] \tag{17}$$

Wherein

$$X = 3x\ C_{D_1}\ \rho_g/16\ D\ \rho_D + 1 \tag{18}$$

The extent of acceleration of drops attained at the end of any throat length may simply be found by settling $x = \ell_t$ in (18) and obtaining u_t as a function of X_t from (17). For example in order for $u_t = 0.90$, $X_t = 1.74$.

For a set of representative conditions, we may explore the value of u_t. Taking air and water at°20 C, $\rho_g = 1.205 \times 10^3$ gm/cm^3, $\rho_D = 1.0$ gm/cm^3 D = 100 microns, with $Re_{D_1} = 400$, $C_{D_1} = 0.608$, and $\ell_t = 1$ ft:

$$X_t = \frac{3 \times (12 \times 2.54)\ 0.608 \times 1.205 \times 10^{-3}}{16 \times (100 \times 10^{-4}) \times 1.0} + 1 = 1.42$$

and $u_t = 0.83$. The gas velocity in the throat here is $v_t = 197$ ft/s, a typical value.

In case the drop velocity at the point of introduction cannot be taken as zero (16) would be written:

$$\frac{1}{(v_t - v_{D_1})^{0.5}} \int_{V_{D_1}}^{V_D} \frac{v_D \, dv_D}{(v_t - v_D)^{1.5}} = \frac{3}{4} \frac{\rho_g}{\rho_D} \frac{C_{D1}}{D} \int_0^x dx \tag{19}$$

These results cannot be applied outside the throat because of the limitations placed upon the conditions of integration. To cover the cases where drops are introduced ahead of the throat, types a(iv), b(ii), where $z_1 < z_2$, integration would be done in two steps. First from z_1 to z_2, taking v_{D1} at z_1 and using (11a) to represent v_g as a function of z between z_1 and z_2. This would determine v_{D2} at z_2, which would become the starting point of the second integration. This would be through the throat, as above, from z_2 to z_3 except that $v_{D2} \neq 0$, and (15) (or (19)) would become:

$$\frac{1}{(v_t - v_{D_2})^{0.5}} \int_{v_{D2}}^{v_D} \frac{v_D \, dv_D}{(v_t - v_D)^{1.5}} = \frac{3}{4} \frac{\rho_g}{\rho_D} \frac{C_{D2}}{D} \int_{z_2}^{z} dx \tag{20}$$

In case it is desired to carry the integration into the diverging section, still another calculation would have to be done using (11b) to relate v_g and z beyond z_3. The lower limits of the integration would be $z = z_3$ and $v_D = u_t v_t$.

Hollands and Goel [8] have considered these possibilities and applied them to the question of calculating pressure drop across the venturi, to be discussed below. The equations must be solved by numerical integration and the results are presented in a series of non-dimensionalized charts.

The above modifications, in principle, would also be applied to the rod bank, type b(iii), using the circular profile of the rods to describe the converging and diverging sections.

All of the above treatment is based upon the equation of motion applied to individual drops. Where drops are flowing in a group, as it were, their behavior may be modified, especially if the drop concentration is high. Crowe [10] studied the acceleration of solid spherical particles fairly uniformly distributed across the entire cross section of a shock tube. He obtained values of C_{DA} which lie parallel to and just above the "standard" curve for C_D in the range of Re_1 for 400 to above 1000. Hesketh, Engel, and Calvert [11] studied acceleration of droplets and found that a "cloud" type of atomization could occur where water is injected in a stream from nozzles greater than about 1 mm in diameter. Particles making up the clouds were much finer and accelerated more slowly than would otherwise be expected.

3. Drop Size and Size Distribution

a. Gas Atomized Sprays. The size of drops is determined by the shattering impact of the high speed gas. This is governed by a critical value of a dimensionless group called the Weber number:

$$We = \rho_g v_t^2 D/\sigma \tag{21}$$

This number expresses the magnitude of the ratio of the inertial stress produced by the gas, to the surface tension stress opposing deformation.

Values of the We_{crit} have been reported to range from 6-11 for various liquids, when the stress is applied slowly. Radhakrishnan and Licht [12] found that a value of We = 5 fitted the experimental data of Boll et al [13] where D is the Sauter mean diameter, at conditions in a type b(i) venturi: $L/G = 10 - 20$ gal/M ft^3, $v_t = 150 - 300$ ft/s.

Boll and his group [13] measured the Sauter mean diameter of drops produced by injection through a series of nozzles located 12" upstream of the throat of a rectangular section venturi. The value of r_t = 7", ℓ_t = 12", throat width = 12", $\beta = 12\frac{1}{2}°$ converging, $\beta = 3\frac{1}{2}°$ diverging. This represents a one foot slice taken out of a full-scale venturi in which the width would be several feet. Their results were correlated by

$$\bar{D}_{1,2} = [283{,}000 + 793\ (L/G)^{1.932}]/v_{g_1}^{\ 1.602} \tag{22}$$

where $\bar{D}_{1,2}$ (see equation (15), Chapter 2), is in microns and v_g, in ft/s. L/G ranged from 4.5-18 gal/M ft^3, v_t from 100-300 ft/s., and $v_{g1} = 0.725\ v_t$ from (11a).

No data have been found for types b(ii) and b(iii) venturis, nor any data on size-distribution for any of the type b's. Some authors have used data from studies on preformed sprays (as reported below), but conditions are not comparable [12] and it seems to be a very questionable practice. Estimates of $\bar{D}_{1,2}$ may be made using We_{crit} of 6-10 in these cases.

b. Pre-formed Sprays. This term is taken to mean sprays produced from atomizing nozzles introduced with the gas stream ahead of the throat, as in type a(iv). The nozzles may or may not be pneumatic. Such sprays and nozzles may also be used in gravity chambers.

(1) Pressure nozzles. Liquid under high pressure is forced into a whirl chamber through a tangential passage (hollow-cone) or directly through an orifice (full-cone). Experiments by Nelson and Stevens [14] on drop-sizes produced have been correlated in terms of the volume-median (same as MMD) diameter $\tilde{x}$ by defining

$$Y = \log_{10} \tilde{x}/D_o \tag{23}$$

and

$$Z = \log_{10} Re_o\ (We_o/Re_o)^a\ (u_t/u_a)^b \tag{24}$$

Then

$$Y = AZ^2 + BZ + C \tag{25}$$

In these dimensionless expressions: D_o = diameter of nozzle orifice; u_a x = Q_ℓ/A = average axial liquid velocity through orifice; $u_t = u_a$ x tan $\theta/2$ = average tangential velocity through orifice; θ = maximum cone angle of spray; $Re_o = D_o u_a \rho_D/\mu_D$; $We_o = u_a^2 D_o \rho_g/\sigma$. For water sprays: a = 0.2; b = 1.2; A = -0.144; B = 0.702; C = -1.260. For organic liquids: a = 0.55; b - 1.2; A = -0.0811; B = 0.124; C = -0.186.

Drop-size distribution has been found to fit the upper-limit function of Mugele and Evans [17], as well as the Rosin-Rammler distribution [15], both of which are discussed in Chapter 2. However, a square-root normal distribution is also useful here [14, 16]. This consists of using

$$A = \log_{10} s(We_o/D_o)^{0.5} \tag{26}$$

$$B = \log_{10} We_o/\sqrt{Re_o} \tag{27}$$

and

$$A = 0.150 B^2 - 0.359 B + 0.986 \tag{28}$$

They also give a graph from which the Sauter mean drop size can be determined from s and $\tilde{x}$.

(2) Pneumatic nozzles. The usual design consists of a central core of liquid surrounded by a concentric converging nozzle for air. In addition to the geometry and dimensions of the nozzle, the relative velocity of air to liquid at the nozzle and the ratio of the mass flow rates are basic parameters which govern the nozzle performance. This flow rate ratio in the nozzle must not be confused with the L/G ratio in the venturi when such nozzles are used.

A classical study by Nukiyama and Tanasawa [18] is often cited for performance of these nozzles. Their correlation for the Sauter mean drop diameter produced is

$$\bar{D}_{1.2} = \frac{585}{v_{\overline{rel}}}\sqrt{\frac{\sigma}{\rho_D}} + 597\left[\frac{\mu_D}{\sqrt{\sigma\rho_D}}\right]^{0.45}\left(\frac{1000\ Q_\ell}{Qg}\right)^{1.5} \tag{29}$$

where $\bar{D}_{1,2}$ = Sauter mean drop diameter, microns; $v_{rel} = (v_a - v_\ell)$, v_a = linear velocity of liquid, m/s; σ, dynes/cm; ρ_D, gm/cm^3; μ_D, poises; Q_ℓ, m^3/min; Q_g, m^3/min. These studies covered nozzles ranging from 0.2 to 1.0 mm diameter for liquid and 1-5 mm for air. Relative velocities ranged from 260 ft/s to sonic, Q_ℓ/Q_g from 0.85 to 15 in gal/M ft . The liquids studied were gasoline, water, alcohol and heavy oils.

A more recent study by Kim and Marshall [19] employed an improved technique for measuring the drop sizes. For a single-air nozzle they found

$$D_m = 160.8\,\frac{\sigma^{0.41}\mu_D^{0.32}}{(v_{rel}^2\,\rho_g)^{0.57}\,A^{0.36}\,\rho_D^{0.16}} + 1573\,\frac{\mu_D^2}{(\rho_D\,\sigma)^{0.17}}\left(\frac{Mg}{M_\ell}\right)^n\frac{1}{v_{rel}^{0.54}} \tag{30}$$

where D_m = mass median drop diameter, microns; A = cross-sectional area for air flow, cm^2; M_g = atomizing air mass flow rate, g/s; M_ℓ = liquid mass flow rate, g/s, and the other symbols and units are the same as for (29). Here n = -1 if $M_g/M_\ell < 3$, and n = -0.5 if $M_g/M_\ell > 3$. These experiments included nozzles ranging from 1.83-2.54 mm diameter for liquid and 3.05-6.9 mm for air. Relative velocities ranged from 250 ft/s to sonic and relative flow rates from 0.75 to 75 in gal/M ft^3. The liquids were limited to molten wax and wax-polyethylene mixtures.

The drop size distribution did not follow the log-normal, square-root-normal, or Rosin-Rammler forms. Kim and Marshall correlated it by

$$\Phi_v = \frac{1.15}{1 + 6.67\exp(-2.18\ D/D_m)} - 0.150 \tag{31}$$

where Φ_V = volume fraction-less-than size D, and $\bar{D}_{1,2} = 0.83\ D_m$. Licht [20] found that the upper-limit function fitted well with a maximum drop size always equal to 2.97 times the mass median, and expressed as

$$\Phi_v = 1/2 \left[1 + \text{erf}\ \frac{\ln u/0.52}{\sqrt{2}\ \ln 2.94}\right] \tag{32}$$

where $u = D/D_{max}$.

For corresponding values of L/G and v_{rel}, applied to the air-water system, (30) leads to $\bar{D}_{1,2}$ values which are less than half of those obtained from (29).

4. Particle Capture Efficiency

The model for capture efficiency of the stream of accelerating drops is developed according to the same principles as have been used previously for spray chambers. A material balance over a control volume of length dx, taken anywhere downstream of the point of drop introduction, for the particulate removed by collision with drops is:

$$-v_g A\ dn_{\rho_i} = \eta_{T_i} \left(\frac{\pi D^2}{4}\right) (v_g - v_D) n_{p_i}\ n_D\ A\ dx \tag{33}$$

where n_{pi} = number of particles of size d_{pi} per unit volume of gas n_D = number of drops (targets) per unit volume of venturi. This assures that both particles and drops are uniformly distributed over the cross-sectional area of the venturi, and that each drop collects particles independently and in the same manner as every other drop of the same size.

Assuming further that no drops are lost either by coalescence with each other or by transfer to the duct wall, the value of n_D is given by

$$n_D = \frac{\text{number of drops in control volume at any instant}}{\text{volume of gas in control volume at any instant}}$$

$$n_D = \frac{M_\ell}{v_D \rho_D} \frac{6}{\pi D^3} \Bigg/ \frac{M_g}{v_g \rho_g} = \frac{6}{\pi} \frac{Q_\ell}{Q_g} \frac{v_g}{v_D D^3} \tag{34}$$

Combining (33) and (34) leads to

$$-\frac{dn_{p_i}}{n_{p_i}} = \frac{3}{2} \frac{\eta_{T_i}}{D} \frac{Q_\ell}{Q_g} \frac{(v_g - v_D)}{v_D} dx \tag{35}$$

which must be integrated along the axis of the venturi in order to obtain the grade efficiency of collection of particles d_{pi} by drops of size D. This integration depends upon the interrelationship among v_g, v_D, and x based upon (14) as discussed above. The exact nature of it will vary according to the conditions of drop introduction and venturi geometry. A comprehensive solution has not been carried out but several special cases are available and useful.

The simplest case uses an approximation to Ingebo's data [6] for C_{DA} in (14), combined with (35). The approximation is $C_{DA} = 55/Re_D$, so that

$$dv_D = \frac{165\,\mu}{4\,\rho_D D^2} \frac{(v_g - v_D)}{v_D} dx \tag{36}$$

and (35) becomes

$$-\frac{dn_{p_i}}{n_{p_i}} = \frac{2}{55} \eta_{T_i} \frac{Q_\ell}{Q_g} \frac{\rho_D D}{\mu} dv_D \tag{37}$$

This may be integrated for the case where $x = 0$ at $z = z_1$, $v_{D1} = 0$ and $x \leqq x_t = z_3$. Equation (36) gives

$$165\,\mu x / 4\,\rho_D D^2 v_t = \ell n\ 1/(1-u) - u \tag{38}$$

where $u = v_D/v_t$, as is analogous with (17). If we may assume η_{T_i} is constant along the throat (which is not a very good assumption), then (37) gives

$$- \ln \frac{n_{p3}}{n_{p1}} = \frac{2}{55} \eta_{T_i} \frac{Q_\ell}{Q_g} \frac{\rho_D D v_t}{\mu} u \tag{39}$$

Throat length and grade-efficiency will be linked by (38) coupled with (39), at $x = x_t$.

A more realistic result may be obtained by allowing for the variation of η_{T_i} along the throat. The basic relationship for η_{T_i} is the same as that used for spherical targets in the analysis of spray chambers above, namely

$$\eta_{T_i} = \left[\frac{Stk_i}{Stk_i + 0.35} \right]^2 \tag{40}$$

when

$$Stk_i = \frac{C_i \rho_{pi} d_{pi}^{\ 2}}{18\mu} \frac{(v_t - v_D)}{D} = \frac{C_i \rho_{pi} d_{pi}^{\ 2} v_t (1-u)}{18\,\mu D} = K_i (1-u)$$

These relationships are substituted into (37)

$$- \frac{dn_{p_i}}{n_{p_i}} = \frac{2}{55} \frac{Q_\ell}{Q_g} \frac{\rho_D D}{\mu} v_t K_i^2 \frac{(1-u)^2}{[K_i(1-u) + 0.35]^2} du \tag{41}$$

This may be integrated for the same conditions as before, namely $u = 0$ at $z_1 = z_2$, or for other appropriate limits. In case diffusiophoresis and/or thermophoresis is contributing to the capture process, this relationship (40) will have to be modified along the lines described in Chapter 5.

Calvert [1] has produced an integrated version of (41) which allows for reconciliation with actual performance for grade penetration.

He assumed that particle collection did not begin until the drops had attained some relative velocity $v_t - v_{D1} = f\ v_t = (1 - u_1)\ v_t$ which should fix the lower limit of integration of (41) at $u_1 = 1 - f$. Further he assumed that acceleration was complete at the end of the throat, i.e., $u_t = 1$, which becomes the upper limit. He expressed the result in terms of f as follows:

$$P_i = \exp\left[\frac{2}{55}\frac{Q_\ell}{Q_g}\frac{\rho_D D}{\mu_g} v_t\ F(K_{pt_i} f)\right] \tag{42}$$

Here

$$K_{pt_i} = \frac{d_{pa_i}^2 v_t}{9\mu D} = \frac{2\ \mathrm{Stk}_i}{(1-u)}$$

$$d_{pa_i}^{\ 2} = C_i \rho_p d_{p_i}^{\ 2} = (\text{aerodynamic diameter})^2$$

and

$$F(K_{pt_i} f) = \frac{1}{K_{pt_i}}\left[-0.7 - K_{pt_i} f + 1.4\ \ell n\left(\frac{K_{pt_i} f + 0.7}{0.7}\right) + \frac{0.49}{0.7 + K_{pt_i} f}\right] \tag{43}$$

Regarding the factor f, Calvert [21] states

> The factor f is an empirical factor which absorbs the influence of various parameters not explicitly included in (42). These parameters include collection by means other than impaction, particle growth due to condensation or other effects, drop sizes other than those predicted, loss of liquid to the venturi walls, maldistribution, and other effects.

Available data indicate that: $f \approx 0.25$, for hydrophobic particles; $f \approx 0.4\text{-}0.5$, for hydrophillic particles; $f \approx 0.5$, for large scrubber tests; and that f increases for $L/G < 0.2\ \ell/m^3$, or 1.9 gal/M ft^3. From the strictly mathematical point of view values of f must be interpreted in terms of u. Thus if $f = 0.25$, $u_1 = 0.75$, which means collection occurs while the drops accelerate from $0.75\ v_t$ up to v_t.

In using (42) and (43), D is taken as $\bar{D}_{1,2}$ from the Nukiyama and Tanasawa relation (29), in which Q_ℓ/Q_g is taken for the venturi as a whole and v_{rel} is based upon velocities at z_2. It has been pointed out above that this is not a proper application for gas-atomized sprays. Furthermore, it ignores the size distribution of drops which should be taken into account in the Stk values.

For the air and water system the Scrubber Handbook [1] presents a number of graphs showing P_i as a function of d_{pa}, for various values of Q_ℓ/Q_g and v_t. From these are derived other graphs giving the aerodynamic cut diameter (value of d_{pa} for which $P_i = 0.50$) as a function of Q_ℓ/Q_g, and v_t. All of these graphs are prepared for $f = 0.25$.

Calvert's design method consists in finding (with the aid of such charts) a set of operating conditions which will yield an aerodynamic cut diameter equal to the required performance cut diameter. The latter is determined from a knowledge of the inlet particle size distribution and the required overall penetration (efficiency) according to the methods described in Chapter 3. The equations, as stated above, do not specifically link the throat length with the performance. In other words an integrated form of (14) or (36) is not used.

A more elaborate but perhaps more accurate approach is to combine (16) with (35), obtaining

$$-\frac{dn_{pi}}{n_{pi}} = 2\,\frac{\eta_{T_i}}{C_{D_1}}\,\frac{Q_\ell}{Q_g}\,\frac{\rho_D}{\rho_g}\,\frac{1}{v_t^{0.5}(v_t - v_D)^{0.5}}\,dv_D \tag{44}$$

Using (40) again for η_{T_i} the form to be integrated becomes

$$\frac{dn_{p_i}}{n_{p_i}} = \frac{2\,Q_\ell\,\rho_D\,K_i^2}{C_{D_1}\,Q_g\,\rho_g}\,\frac{(1-u)^{1.5}}{[K_i(1-u) + 0.35]^2}\,du \tag{45}$$

This has been integrated by Yung et al [24] over the length of the throat only, to give

$$\frac{\ln Pi}{B} = \frac{1}{K_{i1}(1 - u_\ell) + 0.7}\Bigg[4K_{i1}(1 - u_\ell)^{1.5} + 4.2(1 - u_\ell)^{0.5} - 5.02\,K_{i1}^{0.5}\left(1 - u_\ell + \frac{0.7}{K_{i1}}\right)\tan^{-1}\left(\frac{(1 - u_\ell)K_{i1}}{0.7}\right)^{0.5}\Bigg] - \frac{1}{K_{i1} + 0.7}\Bigg[4K_{i1} + 4.2 - 5.02\,K_{i1}\left(1 + \frac{0.7}{K_{i1}}\right)\tan^{-1}\left(\frac{K_{i1}}{0.7}\right)^{0.5}\Bigg] \tag{46}$$

In this expression u_1 is evaluated by (17) using $x = \ell_t$ in X, and K_{i1} is evaluated similarly to (40): $K_{i1} = C_i\rho_p d_{Pi}^{\,2}(v_t - v_{Di})/9\,\mu D$; and $B = Q_\ell\rho_\ell/Q_g\rho_g C_{D1}$. It is assumed that drops are introduced at the throat entrance (z_2) and v_{D1} may if desired be taken as zero. Note that if the drops are fully accelerated in the throat, $u_\ell = 1$, only the second term of (46) is needed. Equation (46) gave good agreement [24] with experimental tests as large scale venturis, using the Sauter mean diameter $D_{1,2}$ as given by (29) for D in K_{i1}.

Boll [7] carried out the integration of (35) numerically without attempting to make any of the simplifications mentioned. He used (14a) for the representation of the drag coefficient, and Dorsch's data [11, Chap. 5] for η_{Ti}, each of which was represented by a series of linear approximations. He also used (29) to estimate $\bar{D}_{1,2}$. Stepwise integration of (35) generated grade efficiency relationships for any given set of operating conditions. These could then be compared with experimental results. Good agreement was obtained in some cases. In others disagreement could be explained as due either to the presence of condensation on drops, or to maldistribution of liquid. He concluded that the data did not contradict the model, as used, which could therefore be applied to study of optimum design and operating conditions. As a design procedure, however, this approach is rather cumbersome.

The use of (46), while also somewhat cumbersome, can be programmed and seems to be better than Boll's procedure. Yung et al [24] have provided a chart for the dimensionless correlation of P_i with K_{i1} and X, which will assist in calculations.

None of the above equations for grade efficiency attempts to take into account the size distribution of the drops. All assume, without proof, that the Sauter mean diameter $\bar{D}_{1,2}$ is a satisfactory representation of the array of drop sizes. There really is no data available on the drop size distribution existing in operating Venturis. In a recent study, Chongvisal [25] explored the effect of assuming various reasonable size distributions and found that better agreement with the experimental overall penetration could be obtained than by using the Sauter mean. The necessary calculations are cumbersome, however.

5. Pressure Drop

a. Inlet Velocity Heads. A momentum balance is written over the control volume of length dx, in a manner analogous to the material balance (33).

$$A dp + M_g\, dv_g + M_\ell dv_D + (M_g + M_\ell) \frac{f_t v_g^2}{2D_e}\, dx = 0 \qquad (47)$$

where the last term represents wall friction based upon equivalent diameter D_e of the venturi, and upon turbulent friction factor f_t. To obtain the total pressure drop, (47) should be integrated from z_0 to z_4. However the estimation of the net loss Δp can be simplified by applying (47) to the throat section only ($dv_g = 0$) and neglecting the wall friction, which is in any case offset by the pressure recovery in the diverging section.

Yung, Barbarika, and Calvert [9] have taken this approach using (47) transformed to:

$$dp = -\rho_D v_t (Q_\ell / Q_g)\, dv_D \qquad (48)$$

For the usual conditions: $v_{D1} = 0$ at $z_1 = z_2$, and v_{D2} at $x = \ell_t$

$$\Delta p = p_3 - p_1 = - \rho_D v_t (Q_\ell / Q_g) v_{D_2} \tag{49}$$

or

$$\Delta p = - \rho_D v_t^2 (Q_\ell / Q_g) u_t$$

Combining this with (17) and (18), leads to

$$\Delta p = - 2\rho_D v_t \; (Q_\ell / Q_g) \left[1 - X^2 + X \sqrt{X^2 - 1} \right] \tag{50}$$

This may be written in the form of N_H (number of inlet velocity heads) lost, as in Chapter 3, by dividing:

$$N_H = \frac{\Delta p}{\rho_g v_t^2/2} = 4 \frac{\rho_D Q_\ell}{\rho_g Q_g} \left[1 - X^2 + X \sqrt{X^2 - 1} \right] \tag{51}$$

Note that as X increases the term in brackets will never exceed 0.5, i.e., no matter how long the venturi throat becomes. This equation was tested [9] against experimental data of Boll [7] using again the Nukiyama and Tanasawa expression (29) for $\bar{D}_{1,2}$ to give D in calculating X from (18). It was shown to fit quite well and is recommended for use.

For example, in an air-water system $\rho_D/\rho_g = 830$, $\mu = 1.81 \times 10^{-4}$ gm/cm s with throat cross-section (35.6 x 30.5) cm, $\ell_t = 30.5$ cm, $v_t = 91.5$ m/s, $Q_\ell/Q_g = 1.0 \ell/m^3$;

$$\bar{D}_{1,2} = \frac{585}{91.5}\sqrt{\frac{70}{1}} + 597 \left[\frac{0.01}{70 \times 1}\right]^{0.45} \left[1.0\right]^{1.5} = 82.4 \ \mu m$$

$$Re_{D_1} = \frac{82.4 \times 10^{-4} \times 91.5 \times 10^2 \times 1.205 \times 10^{-3}}{1.814 \times 10^{-4}} = 500; \; C_{D_1} = 0.55$$

$$X = \frac{3 \times 30.5 \times 0.55}{16 \times 82.4 \times 10^{-4} \times 830} + 1 = 1.46$$

$$N_H = 4 \times 830 \times 1.0 \times 10^{-3} \left[1 - (1.46)^2 + 1.46\sqrt{(1.46)^2 - 1}\right] = 1.40$$

Note that N_H is not quite linear in Q_ℓ/Q_g as this ratio enters into the determination of $\bar{D}_{1,2}$ which is involved in C_{D1} and in X. Note also that N_H is dependent upon v_t as well, as it enters independently into $\bar{D}_{1,2}$ and Re_{D1}, hence into C_{D1}. For example, in the above case when $v_t = 45.8$ m/s, the value of N_H drops to 1.30. If longer throat values ℓ_t are considered, note that the maximum value of N_H will be 1.66.

Boll [7] used the complete equation (47) and performed a numerical integration, step-by-step in order to construct the entire curve of pressure vs. distance x downstream from z_1. This reproduced the experimental pressure measurements very well, and may be carried out when a complete detailed study of pressure variation is desired.

As is the case in the grade-efficiency calculations discussed above, these pressure-drop calculations also ignore the effect of drop-size distribution and use only the Sauter mean size to represent the array. Chongvisal [25] also studied this question and showed that by assuming certain reasonable size distributions (in lieu of actual data) better agreement with experimental pressure-drop values could be obtained. It would seem that a mean diameter based upon mass would be more appropriate than the Sauter mean as the pressure-drop depends upon a momentum transfer to the accelerating drops.

b. Relationship Between Power and Efficiency. Since pressure drop is seen to depend primarily on the energy expended to accelerate the drops, and since this relative motion between drops and gas provides the principal capture mechanism, it is evident that pressure drop and overall collection should be closely linked. As was shown in Chapter 3,

Δp represents power consumption per unit of volumetric gas flow rate Q_g. A relationship between power consumption and overall collection efficiency has been recognized empirically for a long time [22], well in advance of the theoretical developments just described. It forms the basis of another design method, based upon the concept of "contacting power," due to Semrau [23].

The basic relationship may be developed by combining (35) with (47), using (14), for the usual case of drops introduced at the beginning of the throat. Then

$$\frac{dp^{\cdot}}{d\ln n_{p_i}} = \frac{\rho_g v_t}{2} \frac{C_{DA}}{\eta_{T_i}} (v_g - v_D) \tag{52}$$

This might be integrated, for example, using (14a) or (36) for C_{DA} and (40) for η_{T_i} to give

$$\frac{dp}{d\ell n\ n_{p_i}} = C_{D_1} \frac{\rho_g v_t^2}{2} \frac{[K_i(\ -u) + 0.35]^2}{K_i^2(1 - u)^{1.5}} \tag{53}$$

The integration would have to be done numerically because of the complex relation between $\ell n\ n_{p_i}$ and u (such as given by (42) or (46)), but it is clear that the kinetic energy of the gas in the throat $\rho_g v_t^2/2$ is directly involved.

The term "contacting power," as defined by Semrau [23], refers to to power, per unit of Q_g, that is dissipated in contacting gas and liquid, and is ultimately converted into heat. In general it may include three parts: gas-phase contacting power P_g (due to the effective friction loss as measured by Δp), liquid-phase contacting power P_ℓ (due to the pumping of the liquid stream), and mechanical contacting power P_m (due to a mechanically driven rotor, not involved in venturis). Total contacting power is the sum of these: $P_T = P_\ell + P_g + P_m$. The interrelationships are as follows:

	Δp - in. H_2O	Δp - cm H_2O
Pg:	$0.1575\ \Delta p$ hp/Mft^3/min	$0.02724\ \Delta p$ kWh/Mm^3
P_ℓ:	$0.583\ p_f(L/G)$hp/Mft^3/min	$0.02815\ p_f(Q_\ell/Q_g)$Mm^3
P_f:	lb/in^2	atm
L/G:	gal/ft^3	Q_ℓ/Q_g: L/m^3

where P_f = liquid feed pressure.

The empirical relationship is to the effect that

$$\text{N.T.U.} = \alpha(P_T)^\gamma \tag{54}$$

where NTU = number of transfer units overall, as defined in Chapter 3, and α, γ are emiprical constants that depend upon the specific aerosol collected. They are little affected by the scrubber size or geometry, or the manner of applying the contacting power. Note that α and γ automatically take into account the particle size distribution of the specific aerosol, so that no grade-efficiency relation is developed explicitly. Values for α and γ are determined by running pilot scale tests on the actual gas stream, for which portable small size scrubbers may be used conveniently. Tabulations and graphs of these data are available [23].

C. Design Procedures

1. The System

The interaction of the independent design (venturi geometry, liquid introduction) and operating (L/G ratio, gas velocity v_t) parameters is shown in Fig. 3. The mathematical interrelationships have been detailed above, at least in principal. It is evident that the system is sufficiently complex that no comprehensive design method of calculation can be set up without making simplifying approximations, or without resorting to fairly elaborate computerized routines.

Usually a particular type of venturi, and liquid introduction system, is assumed. Then for this situation, combinations of L/G and v_t are determined which will produce the required overall effi-

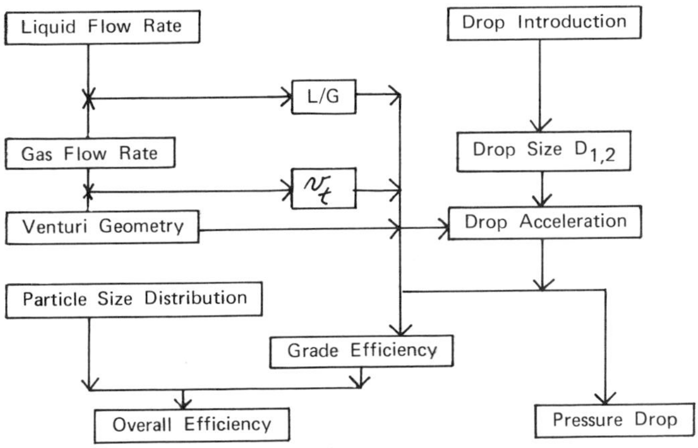

FIG. 3. System analysis for Venturi scrubber.

ciency. As usual there will be a range of these combinations which will "work." For each, the corresponding pressure drop and energy requirements will also be determined. An optimum set of conditions may then be found. Finally, the dimensions of the venturi throat may be calculated.

On the basis of illustrating what may be done by relatively simple methods, the Calvert method will be applied to an example case.

Example 2. The author is indebted to Dr. R. L. Byers for proposing this problem, and for some of the calculations.

Consider a dust stream which is the outlet from a cyclone (primary collector) and is to be collected with at least 99.5% overall efficiency by a secondary collector, possibly a venturi scrubber. The design of such a scrubber is to be investigated. Operating conditions will be: gas flow rate: Q_g = 13,300 acfm; gas temperature: 20°C (same as water temperature); gas pressure: 1 atm; dust properties: ρ_p = 1.5 gm/cm , MMD = 5.9 μm, σ_g = 2.4, c_o = 2.63 gr/acf; inlet mass rate = 300 lbm/hr; outlet mass rate = (1 - 0.995) x 300 = 1.5 lbm/hr.

A type b(i) venturi will be studied: gas-atomized spray, liquid introduced through tubes at the throat, rectangular cross-section r_t = 6", design as illustrated in Fig. 1. It will be assumed that pilot-scale testing has indicated a value of f = 0.25. When v_t is determined, the width of the throat will be fixed by Q_g, r_t, and v_t. When v_t and L/G are determined the length of the throat will be determined through use of X.

A recommended procedure would be as follows: 1. select possible values of Q_ℓ/Q_g (say 1 L/m^3, or L/G = 9.33 gal/Mft^3) and v_t (say 150 m/s, or 457 ft/s). 2. Calculate the Sauter mean drop size, preferably using (22), but alternatively using (29) simply because this has been used in the evaluation of f. Here $\bar{D}_{1,2}$ = 18.7 μm from (22), and $\bar{D}_{1,2}$ = 62 μm from (29). Note the wide difference. 3. Set up the calculation of Stk_i or K_i for use in (42) in terms of particle sized d_{pi} with the corresponding Cunningham factors C_i. Here, in general

$$K_i = \frac{C_i d_{pi}^2}{\bar{D}_{1,2}} \frac{1.5 \times 15{,}000}{18 \times 1.814 \times 10^{-4}} = 6.891 \times 10^6 \frac{C_i d_{pi}^2}{\bar{D}_{1,2}}$$

and for d_{pi} = MMD = 5.9 μm, C_i = 1.03:

$$K_i = 398 \text{ for } \bar{D}_{1,2} = 62\ \mu\text{m}$$

4. Calculate grade efficiencies for values of d_{pi}, using (43) and (44) with $K_{pti} = 2K_i$, and f = 0.25. Since this is to be based upon Calverts' determination of f, $\bar{D}_{1,2}$ must be taken as 62 μm. As an example, for d_{pi} = MMD, (44) gives

$$F(K_{pt}f) = F(796 \times 0.25) = 0.2412 \text{ and (43) gives}$$

$$P_i = \exp\left[-0.2412 \times \frac{2}{55} \times \frac{0.001 \times 1 \times 62 \times 10^{-4}}{1.814 \times 10^{-4}} \times 15000\right] = 0.0112$$

5. With the grade efficiencies calculated in Step 4, and the particle size distribution of the inlet aerosol, calculate the overall efficiency as explained in Chapter 3. If this exceeds the required

value, (here 99.5%) the Q_ℓ/Q_g and v_t selected are a workable combination. If not new value(s) must be selected and the entire calculation repeated. Obviously this procedure could be quite repetitious and cumbersome. It is also tied to the Calvert integration model (43), (44) in order to make use of what is known about f. One alternative approach, on a more fundamental basis, would be to conduct the step wise numerical integration of (35) as done by Boll [7] and discussed above. This is even more cumbersome, requiring a computer routine.

2. Calvert Method

In case the inlet particle size distribution is log-normal, as in the above example, the calculation procedure outlined may be greatly facilitated. The concepts of "required cut diameter" and "performance cut diameter," together with a set of charts, presented by Calvert [1] are utilized.

The "required cut diameter" d_{P50} is the particle size (in an aerosol of given MMD and σ_g) which must be collected with 50% efficiency in order to produce a specified overall (integrated) efficiency. It is a function of $K_{pt}f$). A series of three charts, Figs. 8.3-5, 6, 7 in [1], for values of σ_g = 2.5, 5, 7.5 respectively are available from which d_{P50} may be found. In the example, σ_g = 2.4; $K_{pt}f$ = 199. For 99.5% collection efficiency required, P = 0.005. Using Fig. 8.2-5 (for σ_g = 2.5) the value of d_{P50}/MMD is read as 0.055. Hence, d_{P50} d_{P50} = 0.055 x 5.9 = 0.32 μm. The corresponding aerodynamic diameter (C_i = 1.51) will be d_{Pa} = 1.51 x 1.5 x 0.32 = 0.48.

The "performance cut diameter" is that particle size which will be collected with 50% efficiency under the operating conditions assumed. Only one chart, Fig. 5.36-13 in [1] is available for this, based upon the air-water system and f = 0.25 only. For other systems, or other values of f, this performance cut diameter could be calculated by (43) following the procedure outlined. In the example case, the chart gives the aerodynamic diameter of 0.47 μm at Q_ℓ/Q_g = 1L/m^3 and v_t = 15,000 cm/s. Note that σ_g and MMD are not involved here.

Where the "performance" aerodynamic cut diameter equals the "required" aerodynamic diameter, the set of values for Q_ℓ/Q_g and v_t is correct, i.e., these conditions will work. The agreement of 0.47 vs. 0.48 in the example is taken to be satisfactory.

For these operating conditions, the cross-sectional area of the throat must be $A = Q_g/v_t = 13{,}300/457 \times 60 = 0.485\ ft^2$. For a rectangular shape of height of 1 ft, this would correspond to a width of only 5.8", or a circular shape if diameter 9.43 inches could be used. The maximum pressure drop (regardless of throat length) estimated from (50) would be

$$\Delta\rho = \frac{-2 \times 1.0 \times (15{,}000)^2(0.001) \times 0.5}{980} = 229\ \text{cm}\ H_2O$$

This is a very high pressure drop. These conditions, therefore, are probably far from optimum and other values of Q_ℓ/Q_g and v_t must be explored. With the Calvert charts this can be done rather quickly.

The problem of determining a precise value of throat length remains unresolved by this approach. As a guideline, ℓ_t would have to be close to that used in the pilot scale tests which gave the value of f. This value should be used to find X and then a more precise estimate of Δp. However, as has been noted Δp is not very sensitive to changes in ℓ_t. For example if $1_t = 1$ ft in the above case, calculations similar to those following (51) above would give X = 1.57 and $\Delta\rho = 199$ cm H_2O.

The basic theoretical difficulty here lies in the fact that the integration of (16) used in (50) to get Δp goes from $v_{D1} = 0$ up to v_{Dt} at end of throat, while the integration of (42) to produce P_i, in the Calvert scheme, goes from $v_{D1} = fv_t$ up to $v_D = v_t$ at the end.

REFERENCES

1. S. Calvert, J. Goldschmid, D. Leith, and D. Mehta, *Scrubber Handbook* NTIS, PB 213-016, 1972.

2. S. Calvert, *Chem. Eng.*, 54 (Aug. 29, 1977).

3. S. Calvert, *Amer. Inst. Chem. Eng. Journ*, 16: 392 (1970).

4. C.J. Stairmand, *Trans. Inst. Chem. Eng'g. (London)*, 28: 130 (1950).

5. L. Cheng, *Ind. Eng. Chem. Process Des. Develop.*, 12: 221 (1973).

6. R. Ingebo, NASA Tech. Note 3762 (1956).

7. R.H. Boll, *Ind. Eng. Chem. Fundam.*, 12: 40 (1973).

8. K.G.T. Hollands and K.C. Goel, *Ind. Eng. Chem. Fundam.*, 14: 16 (1975).

9. S.C. Yung, H.F. Barbarika and S. Calvert, *Journ. Air Poll. Control Assoc.*, 27: 348 (1977).

10. C.T. Crowe, Ph.D. Thesis, University of Michigan, Ann Arbor, Mich., 1961.

11. H.E. Hesketh, A.J. Engel, and S. Calvert, *Atmos. Environ.*, 4: 639 (1970).

12. E. Radhakrishnan and W. Licht, *Amer. Inst. Chem. Eng.*, Symposium Series No. 175, 74: 28 (1978).

13. R.H. Boll, L.R. Flais, P.W. Maurer, and W.L. Thompson, *Journ. Air Poll. Control Assoc.*, 24: 934 (1974).

14. P.A. Nelson and W.F. Stevens, *Amer. Inst. Chem. Eng. Journ.*, 7: 80 (1961).

15. W.R. Marshall, Jr., *Chem. Eng. Prog. Monograph Series 2*, 50 (1974).

16. E.H. Taylor and D.B. Harrison, Jr., *Ind. Eng. Chem.*, 46: 1455 (1954).

17. S. Katta and W.H. Gauvin, *Amer. Inst. Chem. Eng. Journ.*, 21: 143 143 (1975).

18. S. Nukiyama and Y. Tanasawa, *Trans. Soc. Mech. Engrs. (Japan)*, 5: 63 (1939).

19. K.Y. Kim and W.R. Marshall, Jr., *Amer. Inst. Chem. Eng. Journ.*, 17: 575 (1971).

20. W. Licht, *Amer. Inst. Chem. Eng. Journ.*, 20: 595 (1974).

21. S. Calvert, in *Air Pollution (A. Stern, ed.) 3rd Ed., Vol IV:* Chap. 6, Academic Press, New York 1977.

22. C.E. Lapple and H.J. Kamack, *Chem. Eng. Progr.,* 51: 110 (1955).

23. K.T. Semrau, *J. Air Poll. Control Assn.,* 10: 200 (1960); 13: 587 (1963).

24. S.C. Yung, S. Calvert, H.F. Barbarika, and L.E. Sparks, *Envir. Sci. and Tech.* 12: 456 (1978).

25. V. Chongvisal, Ph.D. Dissertation, University of Cincinnati, Dept. of Civil and Environmental Eng. (1979).

INDEX

A

B

C

D

E

P

R

S

T

U

V